Cambridge Tracts in Mathematics
and Mathematical Physics

No. 6

ALGEBRAIC EQUATIONS

ALGEBRAIC EQUATIONS

BY

G. B. MATHEWS, M.A., F.R.S., LL.D.

REVISED BY

W. E. H. BERWICK, Sc.D.

Formerly Fellow of Clare College, Professor of
Mathematics in the University College of
North Wales, Bangor

CAMBRIDGE

AT THE UNIVERSITY PRESS

1930

CAMBRIDGE
UNIVERSITY PRESS

32 Avenue of the Americas, New York NY 10013-2473, USA

Cambridge University Press is part of the University of Cambridge.

It furthers the University's mission by disseminating knowledge in the pursuit of education, learning and research at the highest international levels of excellence.

www.cambridge.org
Information on this title: www.cambridge.org/9781107493612

First edition 1907
Second edition 1915
Third edition 1930
First published 1930
Re-issued 2015

A catalogue record for this publication is available from the British Library

ISBN 978-1-107-49361-2 Paperback

PREFACE

THIS tract is intended to give an account of the theory of equations according to the ideas of Galois. The conspicuous merit of this method is that it analyses, so far as exact algebraical processes permit, the set of roots possessed by any given numerical equation. To appreciate it properly it is necessary to bear constantly in mind the difference between equalities in value and identities or equivalences in form; I hope that this has been made sufficiently clear in the text. The method of Abel has not been discussed, because it is neither so clear nor so precise as that of Galois, and the space thus gained has been filled up with examples and illustrations.

More than to any other treatise, I feel indebted to Professor H. Weber's invaluable *Algebra*, where students who are interested in the arithmetical branch of the subject will find a discussion of various types of equations, which, for lack of space, I have been compelled to omit.

I am obliged to Mr Morris Owen, a student of the University College of North Wales, for helping me by verifying some long calculations which had to be made in connexion with Art. 52.

G. B. M.

BANGOR,
August, 1907.

Now that a reprint has been called for, I have taken the opportunity of inserting the condition that a general quintic may be metacyclic in the field of its coefficients. The discovery and calculation of it are due to my colleague, Mr W. E. H. Berwick.

G. B. M.

BANGOR,
July, 1915.

PREFACE TO THIRD EDITION

IN this edition of Dr Mathews' tract I have adjusted an error in § 51. An account of the resolvents of an irreducible sextic has also been included in §§ 54–59. For this I have drawn freely on a recent paper in the *Proceedings of the London Mathematical Society* and am indebted to the Council for permission to make use of it. A few additional notes and references have also been added.

W. E. H. B.

BANGOR,
November, 1929.

CONTENTS

CHAPTER I

GALOISIAN GROUPS AND RESOLVENTS

1. Suppose that $c_1, c_2, \dots c_n$ form a set of assigned algebraic quantities, and that

$$f(x) = x^n + c_1 x^{n-1} + \dots + c_r x^{n-r} + \dots + c_n.$$

If we can find another set of algebraic quantities $x_1, x_2, \dots x_n$ such that

$$\Sigma x_i = -c_1, \quad \Sigma x_i x_j = c_2, \dots, \quad x_1 x_2 \dots x_n = (-)^n c_n \dots\dots (1),$$

we shall have identically

$$f(x) = (x - x_1)(x - x_2) \dots (x - x_n).$$

Under these circumstances (supposing that the algebra we are using is the ordinary one)

$$f(x) = 0$$

for $x = x_1, x_2, \dots x_n$ and for no other values of x.

Thus every solution of (1) leads to the complete solution of the equation $f(x) = 0$. Conversely the complete solution of $f(x) = 0$ in the form $x = \xi_1, \xi_2, \dots \xi_n$ leads to the complete solution of (1), considered as a system of simultaneous equations, in the form

$$x_1, x_2, \dots x_n = \xi_a, \xi_b, \dots \xi_l,$$

where $\xi_a, \xi_b, \dots \xi_l$ represents, in turn, every permutation of $\xi_1, \xi_2, \dots \xi_n$.

If the values $\xi_1, \xi_2, \dots \xi_n$ are all distinct, $f(x) = 0$ has no multiple roots, and the solutions of the simultaneous equations are all distinct, and are $n!$ in number.

If $f(x) = 0$ has multiple roots, its solution may be made to depend upon an equation without multiple roots. Suppose, for example, that $f(x)$ has a root r of multiplicity a; then the first derived function $f_1(x)$, that is to say df/dx, has a root r of multiplicity $(a-1)$. Hence if $\phi = \mathrm{dv}(f, f_1)$, the highest common factor of f and f_1, the equation

$f/\phi = 0$ has coefficients which are rational functions of $c_1, c_2, \dots c_n$, and its roots are the distinct roots of $f(x)$, each occurring only once. Moreover, if $f_i = d^i f/dx^i$, we can, by finding dv (f_1, f_2), dv (f_2, f_3) and so on, determine by rational operations the exact multiplicity of any repeated root of $f = 0$: hence the complete solution of $f/\phi = 0$ leads to that of $f = 0$. In all that follows it will be assumed that f has no multiple roots.

2. It has been proved in various ways that the roots of $f(x) = 0$ actually exist; that is to say, if real or complex values be assigned, at pleasure, to the coefficients, then there are exactly n determinate real or complex numbers $x_1, x_2, \dots x_n$ such that

$$f(x) = \Pi (x - x_i)$$

for all values of x. Another theorem which will be assumed throughout is that every rational symmetric function of the roots can be expressed as a rational function of the coefficients.

3. What gives special interest to the subject in hand is that the actual determination of the roots of a given equation is a problem which differs in complexity according to the assumptions made with regard to the coefficients, and the value of n. Thus, if $n < 5$, and the coefficients are left arbitrary, it is possible to construct an explicit algebraic function of the coefficients which is a root of the equation. For $n > 4$, this is no longer the case; a fact first proved by Abel, who also perceived the real reason for the limitation, namely, the special properties of the group of permutations of n different things when $n < 5$.

When the coefficients are numerically given, the rational roots, if any exist, can be found by trial, and the values of the irrational ones can be found by approximation. With these processes of approximation, however, we shall not be concerned; our main problem is, in fact, the following:

Given a particular equation with numerical coefficients, it is required to find the simplest set of irrational quantities such that all the roots of the given equation can be expressed as finite rational functions, in an explicit form, of the set of irrationals. What is to be understood by the *simplest* set of auxiliary irrationals will appear as we proceed.

4. Before entering upon the general theory, it will be useful to consider the case of a cubic equation with arbitrary coefficients, and roots α, β, γ. Since the value of $\alpha + \beta + \gamma$ is known, it will be sufficient if we can find the values of two other independent linear functions of the roots. If we take an arbitrary linear function $\alpha + l\beta + m\gamma$, this will, in general, assume six values by the permutation of α, β, γ: these values will be the roots of an equation

$$y^6 + m_1 y^5 + \ldots + m_6 = 0,$$

the coefficients of which are rational in l, m and known quantities. Let us try to make this a quadratic in y^3. Then if ω is a complex cube root of unity, there will be six roots of the form

$$y_1,\ \omega y_1,\ \omega^2 y_1,\ y_2,\ \omega y_2,\ \omega^2 y_2.$$

Assuming, as an identity independent of α, β, γ,

$$\alpha + l\beta + m\gamma = \omega(\beta + l\gamma + m\alpha),$$

we have $l = \omega$, $m = \omega^2$: so that we obtain a function

$$y_1 = \alpha + \omega\beta + \omega^2\gamma,$$

the values of which, when α, β, γ are interchanged, become

$$\begin{aligned} y_2 &= \alpha + \omega^2\beta + \omega\gamma, \\ y_3 &= \omega^2\alpha + \omega\beta + \gamma = \omega^2 y_2, \\ y_4 &= \omega\alpha + \omega^2\beta + \gamma = \omega y_1, \\ y_5 &= \omega\alpha + \beta + \omega^2\gamma = \omega y_2, \\ y_6 &= \omega^2\alpha + \beta + \omega\gamma = \omega^2 y_1. \end{aligned}$$

Consequently

$$y_1^3 + y_2^3 = (\alpha + \omega\beta + \omega^2\gamma)^3 + (\alpha + \omega^2\beta + \omega\gamma)^3 = A,$$

a quantity symmetrical in α, β, γ, and therefore rational in the coefficients of the given cubic; in fact,

$$A = 2\Sigma\alpha^3 - 3\Sigma\alpha^2\beta + 12\alpha\beta\gamma = -2c_1^3 + 9c_1c_2 - 27c_3.$$

Similarly $$y_1 y_2 = \Sigma\alpha^2 - \Sigma\alpha\beta = c_1^2 - 3c_2 = B,$$

another rational function of the coefficients: so that y_1^3, y_2^3 are the roots of the rational equation

$$y^6 - Ay^3 + B^3 = 0.$$

Let $$\theta = \left\{\frac{A + \sqrt{(A^2 - 4B^3)}}{2}\right\}^{\frac{1}{3}}$$

with a fixed determination of the radicals involved. Then we may put

$$\begin{aligned} \alpha + \beta + \gamma &= -c_1, \\ \alpha + \omega\beta + \omega^2\gamma &= \theta, \\ \alpha + \omega^2\beta + \omega\gamma &= B/\theta, \end{aligned}$$

and hence

$$3\alpha = -c_1 + \theta + B/\theta = -c + \theta + \frac{A - \sqrt{(A^2 - 4B^3)}}{2B^2}\theta^2,$$
$$3\beta = -c_1 + \omega^2\theta + \omega B/\theta = \ldots,$$
$$3\gamma = -c_1 + \omega\theta + \omega^2 B/\theta = \ldots.$$

By giving θ all its six values, we obtain all the six permutations of α, β, γ.

It will be noticed that the success of this method depends on finding a power of a linear function of the roots which is a two-valued function of the coefficients; this has been done with the help of an auxiliary number ω which is a root of the rational quadratic $\omega^2 + \omega + 1 = 0$.

In a similar way for the general quartic

$$(\alpha - \beta + \gamma - \delta)^2$$

is a three-valued function of the coefficients, and may be explicitly found by means of an auxiliary rational cubic; after this the solution of the quartic may be completed.

5. If, after the manner of Lagrange, we try to extend this process to a quintic, we take ϵ, a complex fifth root of unity, and form the rational equation satisfied by

$$(x_1 + \epsilon x_2 + \epsilon^2 x_3 + \epsilon^3 x_4 + \epsilon^4 x_5)^5.$$

The degree of this is 24, and it is only in special cases that it can be solved in a manner similar to that which is applicable in the foregoing examples. Thus the method breaks down; at the same time, a generalisation of the process, due to Galois, is of the highest importance in the whole of the theory.

6. Galois begins by considering the rational equation satisfied by the most general linear function of the roots. Let $u_1, u_2, \ldots u_n$ be a set of absolutely undetermined symbols, subject merely to the ordinary algebraic laws of combination; and for the sake of brevity let $n! = \mu$. If we put

$$v_1 = u_1 x_1 + u_2 x_2 + \ldots + u_n x_n = \sum_{i=1}^{i=n} u_i x_i,$$

where $x_1, x_2, \ldots x_n$ are the roots (all different) of $f(x) = 0$, we can obtain from v_1, by interchanging the roots in all possible ways, μ essentially different expressions $v_1, v_2, \ldots v_\mu$

The product

$$\prod_{i=1}^{i=\mu} (v - v_i) = v^\mu + b_1 v^{\mu-1} + \ldots + b_\mu = F(v),$$

where v is a new indeterminate, is an integral function of v with coefficients which are integral and rational in $c_1, c_2, \ldots c_n$ as well as in $u_1, u_2, \ldots u_n$ because $F(v)$ is a symmetrical function of the roots of f.

The equation $F(v)=0$ is called the *complete Galoisian resolvent* of $f(x)=0$. Its discriminant is a rational integral function of $c_1, c_2, \ldots c_n, u_1, u_2, \ldots u_n$, which does not vanish identically: so that we may, if we please, assign numerical values to the parameters $u_1, u_2, \ldots u_n$ without making any two roots of the resolvent equal to each other. In particular, these numerical values may be ordinary real integers.

7. The most important property of F is that *any rational function of the roots of f can be expressed as a rational function of any one of the roots of F.*

Let the given rational function be $\phi(x_1, x_2, \ldots x_n)$, and let

$$\phi_1 (=\phi), \phi_2, \ldots \phi_\mu$$

be the expressions obtained from ϕ by applying the substitutions which derive $v_1, v_2, v_3, \ldots v_\mu$ from v_1. These expressions ϕ_i are not necessarily all different in form; and two which have different forms may have the same value. But it must be remembered that ϕ_i is derived from ϕ_1 by the same permutation which changes v_1 to v_i.

Consider the expression

$$\psi(v) = \left\{\frac{\phi_1}{v-v_1} + \frac{\phi_2}{v-v_2} + \ldots + \frac{\phi_\mu}{v-v_\mu}\right\} F(v);$$

$\psi(v)$ is an integral function of v, in general of degree $(\mu-1)$, but possibly lower, and it is a symmetric function of $x_1, x_2, \ldots x_n$. Hence the coefficients of $\psi(v)$ can be expressed as rational functions of $c_1, c_2, \ldots c_n$; and if, after doing this, we put $v=v_1$, it follows from the above identity that

$$\psi(v_1) = \phi_1 F'(v_1),$$

or

$$\phi_1 = \frac{\psi(v_1)}{F'(v_1)} = R(v_1;\ c_1, c_2, \ldots c_n;\ u_1, u_2, \ldots u_n),$$

where R denotes a rational function of the quantities in the bracket. This equality reduces to an absolute identity if on the right-hand side we replace $v_1, c_1, \ldots c_n$ by their expressions in terms of $x_1, x_2, \ldots x_n, u_1, u_2, \ldots u_n$.

The discriminant of F is

$$\Delta = F'(v_1) F'(v_2) \ldots F'(v_\mu),$$

and the quotient $\Delta/F'(v_1)$ is expressible as a rational integral function of v_1: hence we may also put ϕ into the form

$$\phi = \frac{\psi(v_1) F'(v_2) F'(v_3) \ldots F'(v_\mu)}{\Delta} = \frac{J(v_1)}{\Delta},$$

where $J(v_1)$ is a rational integral function of v_1.

It should be observed also that ϕ_i can be expressed as the same function of v_i that ϕ_1 is of v_1.

Finally, ϕ_1 is expressible as a rational function of *any* root of $F(v)$. Thus if we choose v_i, all we have to do is to replace, in the foregoing proof,

$$v_1, v_2, \ldots v_\mu$$

by

$$s_i(v_1), s_i(v_2), \ldots s_i(v_\mu),$$

where s_i is the perfectly definite substitution which converts v_1 to v_i. In general, ϕ is not the same rational function of v_i as it is of v_1.

8. Several important consequences immediately follow from the theorem just proved. In the first place, we may put $\phi = v_i$, and thus infer that

All the roots of the Galoisian resolvent may be expressed as rational functions of any one of them.

An equation having this property is called a *normal* equation; the Galoisian resolvent is accordingly a normal equation. It must be remembered that the same equation may be normal from one point of view and not from another, if, in the definition, we understand "rational function" to mean "rational function with rational coefficients." By a *field of rationality* we shall understand the aggregate of all the expressions obtainable from a finite set of symbols $t_1, t_2, \ldots t_m$ by a finite set of rational operations; that is to say, all the expressions which can be reduced to the form

$$\frac{\phi(t_1, t_2, \ldots t_m)}{\psi(t_1, t_2, \ldots t_m)},$$

where ϕ, ψ are finite polynomials with ordinary whole numbers for their coefficients. The elements $t_1, t_2, \ldots t_m$ may be partly undetermined parameters, or *umbræ*, partly determinate numbers; those which are numerical may be irrational arithmetically, but are here considered rational in the sense of being given or determined. The simplest field of rationality is that of ordinary rational numbers; this is contained in every other field.

If t_{m+1} is any algebraic number or symbol not contained in the field $(t_1, t_2, \ldots t_m)$, the field $(t_1, t_2, \ldots t_m, t_{m+1})$ is said to be obtained from the former field by the *adjunction* of t_{m+1}: this term is specially employed when t_{m+1} is a numerical quantity.

In the case of the Galoisian resolvent we may say, then, that it is a normal equation in the field

$$(c_1, c_2, \ldots c_n; u_1, u_2, \ldots u_n).$$

9. If, in the theorem of Art. 7, we put $\phi = x_i$, we arrive at the proposition that

Every root of an equation without multiple roots can be expressed as a rational function of any one root of its Galoisian resolvent.

If rational values are given to the parameters $u_1, u_2, \ldots u_n$, the resolvent equation becomes normal in the field $(c_1, c_2, \ldots c_n)$. Moreover if $c_1, c_2, \ldots c_n$ are given, not as symbols, but as actual numbers, the resolvent becomes a definite numerical equation. Unless this equation has multiple roots, it is still true that the knowledge of the value of any one root of the resolvent leads to the complete solution of $f=0$; because to calculate the function $\psi(v)$ of Art. 7 in its rational form it is sufficient to know the *values* of the elementary symmetric functions of $x_1, x_2, \ldots x_n$, and these are given by f.

10. The total resolvent $F(v)$ may or may not be reducible without adjunction; in the second case $f(x)=0$ is said to be an equation without *affection*.

The irreducible factors of the resolvent of an affected equation are all of the same degree.

Let $\psi_1(v)$, $\psi_2(v)$ be any two such factors: let v_1 be any root of $\psi_1(v)=0$, and v_2 any root of $\psi_2(v)=0$. Then (Art. 7) v_2 can be expressed as an integral function, $J(v_1)$, of v_1. If the Tschirnhausen transformation $y=J(x)$ is applied to $\psi_1(x)=0$, we obtain an equation $\chi(y)=0$ of the same degree as $\psi_1=0$ which has a solution $y=v_2$ in common with $\psi_2(y)=0$: hence $\chi(y)$ is divisible by $\psi_2(y)$, and the degree of ψ_1 cannot be less than that of ψ_2. By a similar argument, the degree of ψ_2 cannot be less than that of ψ_1; therefore the degrees must be equal.

If h is the degree of each irreducible factor, we have an identity

$$F(v) = \psi_1(v)\,\psi_2(v) \ldots \psi_m(v),$$

with

$$mh = \mu,$$

so that m and h are conjugate factors of μ.

Every one of the equations $\psi_i(v)=0$ is normal, and they are all Tschirnhausen transformations of any one of them. Each may be called a *primary* resolvent of $f(x)=0$. The knowledge of any one root of a primary resolvent leads to the complete solution of $f(x)=0$.

11. A simple example will help to illustrate the results so far obtained. Let the given equation be

$$x^3 - x^2 + x - 1 = 0,$$

and let a, b, c be used instead of u_1, u_2, u_3.

The complete resolvent is $F = \phi\chi\psi$,

where

$$\phi = (v-a)^2 + (b-c)^2, \quad \chi = (v-b)^2 + (c-a)^2, \quad \psi = (v-c)^2 + (a-b)^2.$$

One root of $\phi = 0$ is $a - bi + ci$, and from this the roots $1, i, -i$ of the original equation are obtained. If we put

$$v_1 = a - bi + ci,$$

then

$$\pm \frac{v_1 - a}{b - c}, \; 1,$$

give the roots of $f=0$ as rational functions of v_1.

12. The reducibility of F shows the existence of asymmetrical functions of $x_1, x_2, \ldots x_n$ which nevertheless have rational values. The coefficients of the terms of a primary resolvent $\psi(v)$, considered as a polynomial in $v, u_1, u_2, \ldots u_n$, are all rational; but when expressed in terms of $x_1, x_2, \ldots x_n$ they cannot all be symmetrical, otherwise every permutation of the roots of f would leave $\psi(v)$ unaltered, and this is not the case.

13. Consider now a primary resolvent

$$\psi_1(v) = (v - v_1)(v - v_2) \ldots (v - v_h).$$

Any one of its roots, say v_i, can be derived from v_1 by a perfectly definite permutation of $x_1, x_2, \ldots x_n$: let this be called s_i. Including the identical substitution s_1, we have in connection with ψ_1 just h substitutions $s_1, s_2, \ldots s_h$. It is a most important theorem that *these substitutions form a group*; that is to say, for every pair of substitutions s_a, s_b (the same or different) we have $s_a s_b = s_c$, where s_c is a definite substitution of the same set.

It follows from Art. 7 that since v_b and v_1 are both roots of $F(v)=0$, there is an integral function $J(v)$ such that

$$s_b(v_1) = v_b = J(v_1).$$

Moreover it appears from the same article that

$$J(v_a) = s_a(v_b) = s_a\{s_b(v_1)\}.$$

But since the equations

$$\psi_1(v)=0, \quad \psi_1\{J(v)\}=0$$

have a common root v_1, and the first is irreducible, while both are rational, each root of the first is a root of the second, and in particular

$$\psi_1\{J(v_a)\}=0;$$

that is to say, $s_a\{s_b(v_1)\}$ is a root of $\psi_1(v)=0$, and is therefore equal in value to $s_c(v_1)$, where s_c is a substitution of the set $s_1, s_2, \ldots s_h$. But this equality in value must also be a coincidence in form, on account of the arbitrary nature of the parameters $u_1, u_2, \ldots u_n$. Hence

$$s_b s_a = s_c,$$

it being understood that $s_b s_a$ means the result of first applying s_b and then applying s_a. In a similar way $s_a s_b = s_d$; but s_d is, in general, different from s_c.

14. If ψ_2 is any other of the primary resolvents, there will, in the same way, be a group of substitutions connected with it. This is, in fact, the same group as the one associated with ψ_1. For suppose that

$$\psi_2(v)=(v-v_{h+1})(v-v_{h+2})\ldots(v-v_{2h}):$$

then v_{h+1} can be expressed in the form

$$v_{h+1}=J(v_1),$$

and by the usual argument it follows that

$$\psi_2=\{v-J(v_1)\}\{v-J(v_2)\}\ldots\{v-J(v_h)\}.$$

The notation may be so arranged that

$$J(v_i)=v_{h+i} \qquad (i=1, 2, \ldots h),$$

and this being so, we conclude that

$$v_{h+i}=s_i(v_{h+1}),$$

because v_{h+i} is derived from v_{h+1} by the change of v_1 into v_i, and the only substitution which does this is s_i.

The group $(s_1, s_2, \ldots s_h)$ is called *the Galoisian group* of the equation $f(x)=0$. If the complete resolvent is irreducible without adjunction, $h=n!$ and the Galoisian group consists of all the permutations of $x_1, x_2, \ldots x_n$.

15. We will now select any one of the primary resolvents, denote it by $\psi(v)$, and call it simply, for the present, *the* resolvent of $f(x)$. Assuming nothing about $f(x)$ except that its coefficients are actually given, $F(v)$ and subsequently $\psi(v)$ can be found by rational operations. The degree of $\psi(v)$ in v at once gives the order of the Galoisian group.

But we can go further than this, and determine, from an examination of ψ, the elements $s_1, s_2, \dots s_h$ which form the group. The notation may be so arranged that

$$\psi = (v - v_1)(v - v_2) \dots (v - v_h),$$

$$v_1 = u_1x_1 + u_2x_2 + \dots + u_nx_n.$$

Now the change of v_1 into v_2 effected by the substitution s_2 may also be effected by a substitution σ_2 operating on the parameters $u_1, u_2, \dots u_n$. For instance, if

$$v_1 = u_1x_1 + u_2x_2 + u_3x_3 + u_4x_4 + u_5x_5 + u_6x_6,$$

$$v_2 = u_1x_2 + u_2x_4 + u_3x_6 + u_4x_1 + u_5x_3 + u_6x_5,$$

then $\qquad s_2 = (x_1x_2x_4)(x_3x_6x_5), \quad \sigma_2 = (u_1u_4u_2)(u_3u_5u_6).$

In general, if s_i contains the cycle $(x_ax_b \dots x_kx_l)$, σ_i contains the cycle $(u_lu_k \dots u_bu_a)$ and there is a one-one correspondence between the substitutions s_i and the substitutions σ_i. If σ_i is applied to $\psi(v)$ in its rational form, the result is a function $\chi(v)$ of the same order, which has a root v_i, and therefore coincides with $\psi(v)$. Thus there are at least h distinct permutations σ, forming a group, which leave $\psi(v)$ formally unaltered. The same argument applies to the other primary resolvents obtained from F, and since there are only hm substitutions σ altogether, it follows that there are precisely h substitutions σ which leave ψ formally unaltered; from each of these we can deduce uniquely a substitution s belonging to the Galoisian group.

For instance, in the example of Art. 11, if we take ψ as the resolvent,

$$\sigma_1 = 1, \quad \sigma_2 = (ab),$$

and the corresponding Galoisian group is

$$s_1 = 1, \quad s_2 = (x_1x_2).$$

After obtaining the elements of the Galoisian group

$$G = (s_1, s_2, \dots s_h),$$

its properties, as a group of substitutions, or more generally as an abstract group, may be investigated. These are, in themselves, wholly independent of the values of $x_1, x_2, \dots x_n$.

16. It will now be supposed that the coefficients of f are numerical; and, as explained in Art. 8, any quantity in the field $(c_1, c_2, \dots c_n)$ will be considered rational, no matter whether the coefficients c_i are arithmetically rational or not. It will now be proved that

Every rational function of the roots of f which is unchanged **in numerical value** *by the substitutions of the Galoisian group has*

a value which can be expressed in a rational form: that is to say, it is equal in value to a certain rational function of the coefficients of f.

Let the given function be $\phi(x_1, x_2, \dots x_n)$ and let $v_1, v_2, \dots v_h$ be the roots of the resolvent $\psi(v)$. Then (Art. 7) there is an integral function $J(v)$ such that

$$\phi = \phi_1 = J(v_1),$$

$$\phi_2 = J(v_2),\ \phi_3 = J(v_3), \dots \phi_h = J(v_h),$$

where $\phi_2, \phi_3, \dots \phi_h$ are derived from ϕ by applying the Galoisian substitutions $s_2, s_3, \dots s_h$. Hence

$$\phi_1 + \phi_2 + \dots + \phi_h = J(v_1) + J(v_2) + \dots + J(v_h)$$
$$= S(c_1, c_2, \dots c_n;\ u_1, u_2, \dots u_n),$$

where S is a rational function, because $\Sigma J(v_i)$ is a symmetrical function of $v_1, v_2, \dots v_h$ and the coefficients of $\psi(v)$ are rational. If, now, $\overline{\phi}_i$ means the value of ϕ_i, we have, by hypothesis,

$$\overline{\phi}_1 = \overline{\phi}_2 = \dots = \overline{\phi}_h = \frac{1}{h}(\overline{\phi}_1 + \overline{\phi}_2 + \dots + \overline{\phi}_h)$$

$$= \frac{1}{h}\overline{S}(c_1, c_2, \dots c_n;\ u_1, u_2, \dots u_n),$$

where $\overline{S}$ means the value of the rational function S.

If the coefficients c_i are represented symbolically, the function S, even in its lowest terms, may contain the parameters explicitly; in this case the value of ϕ is expressible as the quotient of any numerical coefficient in the numerator of S by the corresponding coefficient in the denominator. The fact that we thus have alternative rational equivalents for ϕ implies one or more rational relations connecting the coefficients c_i. If, on the other hand, the coefficients c_i are actually given as numbers in a definite field (for instance, if they are all of the form $\alpha + \beta\sqrt{2}$, with α, β rational numbers in the ordinary sense), the parameters, at the last stage of the process, disappear of themselves, and we obtain the value of ϕ as a definite number in the field. The point of the proof is then that the value in question is expressible as a quantity in that particular field.

17. Conversely, *every rational function of the roots which has a rational value keeps that value when any substitution of the Galoisian group is applied to it.*

Let ϕ be the rational function, and A its rational value. Expressing ϕ as a rational integral function of v_1, we have

$$\phi = J(v_1) = A,$$

and hence the rational equation

$$J(v) - A = 0$$

is satisfied by v_1, and consequently by $v_1, v_2, \dots v_h$.

Thus $$J(v_i) = A;$$

that is to say, $$A = s_i J(v_1) = s_i\phi,$$

which proves the theorem. It must be remembered, of course, that $s_i\phi$ may or may not be formally different from ϕ. Moreover, in any actual case, if we reduce $J(v_1)$ to a degree lower than h by means of $\psi(v_1) = 0$ we shall in the end obtain A explicitly, if the value of ϕ is actually rational: so the process of Art. 7, applied to a particular function ϕ and a particular equation f, decides whether the value of ϕ is rational or not.

Finally, there are rational functions of the roots which have rational values, but change these values when substitutions other than those of G are applied to them.

To show this, let θ be an undetermined rational quantity; then

$$\psi(\theta) \equiv (\theta - v_1)(\theta - v_2)\dots(\theta - v_h) = A,$$

where A is rational in $(\theta;\ c_1, c_2, \dots c_n;\ u_1, u_2, \dots u_n)$. If t is any substitution not contained in the Galoisian group, $t\psi(\theta) = \psi_1(\theta)$, where ψ_1 is a primary resolvent distinct from ψ. Considered as an equation in θ,

$$\psi_1(\theta) - \psi(\theta) = 0$$

cannot have more than $(h-1)$ roots, even when the parameters have fixed numerical values (subject to the usual restriction $\Delta \neq 0$). Since there are $(m-1)$ conjugate resolvents into which ψ can be transformed, we have to exclude at most $(h-1)(m-1)$ values of θ. For any other rational value of θ, it is the substitutions of G, and these alone, which leave the value of $\psi(\theta)$ unaffected.

Every coefficient of ψ, considered as a polynomial in $\theta, u_1, u_2, \dots u_n$, is unaffected in value by the substitutions of G; it not unfrequently happens that some one of these coefficients, or a simple linear combination of them, can be seen to have its value changed by all substitutions not belonging to G; in this case it may be taken instead of $\psi(\theta)$. For an example, see Art. 29 below.

As a result of the three theorems last proved we may define the Galoisian group of f as the aggregate of those permutations of $x_1, x_2, \dots x_n$ which leave unaltered in value *every* rational function of the roots which has a rational value.

18. If ϕ is any rational function of the roots of f it has been proved that ϕ can be expressed as an integral rational function of v_1, and it has been observed that in virtue of $\psi(v_1)=0$, this integral function can be reduced so that its degree does not exceed $(h-1)$. An independent proof of this affords a little more information. If, with the usual notation,

$$\chi(v) \equiv \left\{\frac{\phi_1}{v-v_1}+\frac{\phi_2}{v-v_2}+\ldots+\frac{\phi_h}{v-v_h}\right\}\psi(v),$$

$\chi(v)$ is an integral function of v which is also rational, because it is unaltered by any substitution of G. Consequently

$$\phi=\phi_1=\frac{\chi(v_1)}{\psi'(v_1)},$$

a rational function of v_1, which may also be reduced to the form

$$\phi_1=\frac{\chi(v_1)\,\psi'(v_2)\,\psi'(v_3)\ldots\psi'(v_h)}{\delta}$$
$$=\frac{j(v_1)}{\delta},$$

where δ is the discriminant of ψ, and $j(v_1)$ is an integral function, which in virtue of $\psi(v_1)=0$ may be supposed put into its reduced form, so that its degree is not greater than $(h-1)$. If ϕ is an integral function of the roots, the coefficients of j will be integral in

$$c_1, c_2, \ldots c_n;\ u_1, u_2, \ldots u_n.$$

Similarly, $$\phi_i=j(v_i)/\delta \qquad (i=1, 2, 3, \ldots h).$$

The quantity δ is not zero, because it is a factor of Δ.

The substitutions of G give to ϕ the different forms $\phi_1, \phi_2, \ldots \phi_h$: these, however, need not be all different in value. Those substitutions *of the Galoisian group* which leave ϕ unaltered in value form a subgroup, or factor, of G which may be called the *invariant group of* ϕ.

In fact, if s_a, s_b are any two such substitutions,

$$s_a\phi=s_b\phi=\phi,$$

numerically: hence $$s_a\phi-\phi=0,$$

and since the expression on the left hand is a rational function of the roots which has the rational value 0, we may, by Art. 17, apply the substitution s_b to it, and conclude that

$$s_b(s_a\phi-\phi)=0;$$

that is, $$s_b(s_a\phi)=s_b\phi=\phi$$

numerically. Hence s_as_b leaves the value of ϕ unaltered, and the substitutions in question form a group, because $s_a s_b$ is identical with a substitution of G, and it has been shown that it leaves ϕ unaltered in value.

It must be carefully remembered that the invariant group of ϕ consists exclusively of substitutions which belong to G. There may be other substitutions which leave ϕ unaltered in value, or even in form, but if they are not in the Galoisian group they are not to be included. The fact is that we cannot infer for certain that if $s_a\phi - \phi = 0$, then $s_b(s_a\phi - \phi) = 0$, unless s_b belongs to the Galoisian group (cf. Art. 17, end).

Writing, as usual, $s_i\phi = \phi_i$, the function ϕ is a root of the rational equation

$$(\phi - \phi_1)(\phi - \phi_2) \dots (\phi - \phi_h) = 0.$$

But if the invariant group of ϕ is of order $k > 1$, the roots of this equation are repeated each k times : hence if we put $h/k = l$, which is necessarily an integer, ϕ is a root of a rational equation

$$\phi^l + b_1\phi^{l-1} + \dots + b_l = 0.$$

19. *If $f(x)$ is reducible without adjunction, its Galoisian group is intransitive, and conversely.*

First suppose that G is intransitive : this means that a certain number of roots

$$x_1, x_2, \dots x_r \qquad (r < n)$$

are only interchanged among themselves by the substitutions of G. Consequently (Art. 16)

$$(x - x_1)(x - x_2) \dots (x - x_r)$$

being unaltered by any substitution of G has rational coefficients, and $f(x)$ is reducible without adjunction.

Conversely, suppose that $f(x)$ has a rational factor

$$f_1(x) = (x - x_1)(x - x_2) \dots (x - x_r) \qquad (r < n),$$

then, if G is transitive, it must contain a substitution s, which converts some one of the roots $x_1, x_2, \dots x_r$, say x_1, into a root x_{r+1}, formally different from $x_1, x_2, \dots x_r$. Hence sf_1 contains the factor $(x - x_{r+1})$: but since f_1 is rational $sf_1 = f_1$, and consequently

$$0 = f_1(x_{r+1}) = (x_{r+1} - x_1)(x_{r+1} - x_2) \dots (x_{r+1} - x_r);$$

implying that $f(x) = 0$ has equal roots, contrary to hypothesis. Hence if $f(x)$ is reducible, G is intransitive. The example of Art. 11 gives a simple illustration.

It is possible to resolve $f(x)$ into its irreducible factors by means of rational operations, even when the coefficients are connected by known algebraic relations. Unless the contrary is expressed, it will be assumed henceforth that f is irreducible without adjunction.

20. Suppose that by the adjunction of a quantity θ the resolvent ψ becomes reducible in the field $(\theta;\ c_1, c_2, \dots c_n;\ u_1, u_2, \dots u_n)$. If we have

$$\psi = \psi_1 \psi_2,$$

it follows by comparing coefficients that θ satisfies one or more rational equations in the original field. These must be consistent with each other, so that θ must satisfy a definite irreducible equation

$$a(\theta) \equiv \theta^l + a_1\theta^{l-1} + \dots + a_l = 0$$

with rational coefficients, which we may suppose integral because, if necessary, θ may be replaced by $z\theta$, where z is any rational quantity.

If, by any means, this irreducible equation has been found, it is possible to actually resolve ψ into its irreducible factors *in the new field*; and this resolution is unique. We shall have

$$\psi = \chi_1\chi_2 \cdots \chi_i$$

and v_1 will be a root of one of the irreducible equations $\chi_i = 0$. Arranging the notation so that $\chi_1(v_1) = 0$, and for convenience putting $\chi_1 = \chi$, we have an equation

$$\chi(v) = 0,$$

which, in the new field, will serve as a primary resolvent of $f = 0$. This is clear, because $\chi(v)$ is only a transformation of a product

$$(v - v_1)(v - v_a) \dots (v - v_c)$$

so that (Art. 7) $\chi(v) = 0$ is a normal equation; and every rational function of $x_1, x_2, \dots x_n$ can be expressed, in the new field, as an integral function of v_1, the degree of which is less than that of χ, and which is not of higher degree than $(l - 1)$ in θ. As in Art. 10 it can be proved that the functions $\chi_1, \chi_2, \cdots \chi_i$ are all of the same degree in v, and are Tschirnhausen transformations of each other.

In expressing any rational function of $x_1, \dots x_n$ as a reduced function of v_1 in the new field, we may proceed as follows. In the original field let $\phi = j(v_1)$ be the reduced expression for ϕ (Art. 18); divide $j(v)$ by $\chi(v)$ until the remainder is of degree lower than that of χ. We thus obtain an identity

$$j(v) = Q(v)\chi(v) + k(v),$$

and by putting $v = v_1$, we have

$$\phi = j(v_1) = k(v_1),$$

because $\chi(v_1)=0$. The coefficients of k, which are integral functions of θ, may be reduced to their lowest degree by dividing them by $a(\theta)$.

It will be noticed that $\chi(v)$ must contain θ explicitly, because it is a factor of $\psi(v)$, which is irreducible in the old field.

21. We are now approaching the culminating point of Galois's theory. Unless G is a simple group, it will contain self-conjugate factors distinct from the identical substitution: and among these there will be a certain number of maximum self-conjugate factors. Let Γ be a maximum self-conjugate factor of G, of order k and of index $l\,(=h/k)$ with respect to G. The notation may be so arranged that

$$\Gamma=(s_1,\ s_2,\ \dots\ s_k).$$

Let z be an undetermined rational number, and

$$\theta=\phi(x_1,\ x_2,\ \dots\ x_n)\equiv(z-v_1)(z-v_2)\dots(z-v_k),$$

where $v_1, v_2, \dots v_k$ are the roots of the resolvent $\psi=0$, which correspond to the substitutions of Γ. Then the value of θ is unaltered by any substitution of Γ, and by choosing z properly (Art. 17) we can make sure that the value of θ is altered by every substitution of G which is not contained in Γ.

Consequently θ is a function of which Γ is the invariant group, and is a root of a rational equation

$$a(\theta)\equiv\theta^l+a_1\theta^{l-1}+a_2\theta^{l-2}+\dots+a_l=0.$$

So long as $z, u_1, u_2, \dots u_n$ remain undetermined, the coefficients in this equation are integral in the field $(z\,;\ u_1, u_2, \dots u_n\,;\ c_1, c_2, \dots c_n)$: it is possible to give fixed rational integral values to $z, u_1, u_2, \dots u_n$ so as to make the coefficients rational in $(c_1, c_2, \dots c_n)$.

22. It is important to determine the Galoisian group of the equation satisfied by θ. To do this, it is necessary to use a lemma, derived from the elements of the theory of groups. All the substitutions of G may be arranged in the form

$$\begin{array}{ccc} s_1, & s_2, & \dots\ s_k \\ t_2s_1, & t_2s_2, & \dots\ t_2s_k \\ \vdots & \vdots & \vdots \\ t_ls_1, & t_ls_2, & \dots\ t_ls_k \end{array}$$

where $t_2, t_3, \dots t_l$ are distinct elements suitably chosen from G.

If any substitution s of G be applied by premultiplication to the elements of a row in this scheme it will produce a new row which consists either of the elements of the same row, usually in a different order or else the elements of another row, usually in a different order. In no case can elements of the same row be changed into elements of two different rows.

To prove this, suppose, if possible, that, for instance,

$$st_2s_1 = t_a s_p, \quad st_2 s_2 = t_b s_q,$$

where a, b are different. Then, since $s_1, s_2, \dots s_k$ form a group,

$$st_2 = t_a s_p s_1^{-1} = t_a s_i, \quad st_2 = t_b s_q s_2^{-1} = t_b s_j :$$

therefore $$t_a s_i = t_b s_j, \quad t_a = t_b s_j s_i^{-1} = t_b s_r$$

which is impossible, because $t_b s_r$ is in the bth row, and (on account of the way in which $t_1, t_2, \dots t_l$ are chosen) is distinct from t_a, which is in the ath row.

Hence we may say that the application of any substitution of G produces a permutation of the rows of the table. These permutations form a group, denoted by G/Γ, and called the *complementary* group (or factor-group) of G with respect to Γ. The only substitutions of G which leave the first row in its place are the elements of Γ, and these leave every other row in its place, because

$$s_i t_j s_k = t_j s_l s_k = t_j s_m$$

for all values of i, j, k, since Γ is self-conjugate.

Moreover any substitution which converts the first row into the ith must be of the form $t_i s_a$. Applying this to any element $t_i s_b$ of the jth row, we obtain

$$t_i s_a \,.\, t_j s_b.$$

Now because Γ is self-conjugate, we may put

$$s_a t_j = t_j s_c,$$

and hence $$t_i s_a t_j s_b = t_i t_j s_c s_b = t_i t_j s_c'.$$

Finally $t_i t_j = t_k s_d$, where t_k is a definite substitution determined by t_i, t_j alone: hence

$$t_i t_j s_c' = t_k s_d s_c' = t_k s_e$$

and the substitution $t_i s_a$ converts the jth row into the kth. Conversely, the only substitutions which change the jth row into the kth are those which change the first row into the ith. Consequently G/Γ, considered as a group of permutations of rows, may be represented in the form

$$(\tau_1, \tau_2, \dots \tau_l),$$

where τ_i is the definite substitution of G/Γ which changes the first row into the ith.

The substitutions $t_1, t_2, \dots t_l$ do not, as a rule, form a group: but they behave like a group when considered as operations on the rows of the table.

23. It will now be shown that the Galoisian group of the equation $a(\theta)=0$ is holoedrically isomorphic with G/Γ. The values of θ are all different, and we may denote them in such a way that

$$\theta_i = t_i\theta_1 = t_i s_j \theta_1. \qquad \begin{pmatrix} i=1, 2, \dots l \\ j=1, 2, \dots k \end{pmatrix}$$

This being so, every permutation of rows in G/Γ corresponds to a permutation of $(\theta_1, \theta_2, \dots \theta_l)$, and *every* substitution of G produces on $(\theta_1, \theta_2, \dots \theta_l)$ the same permutation as it does in the rows. Now let $Q(\theta_1, \theta_2, \dots \theta_l)$ be any rational function of the roots of $a(\theta)=0$ which has a rational value. Then

$$Q(\theta_1, \theta_2, \dots \theta_l) = R(x_1, x_2, \dots x_n),$$

where R is another rational function. Since the value of R is rational, it is unchanged numerically by any substitution of G. This substitution applied to Q produces a permutation of $\theta_1, \theta_2, \dots \theta_l$ corresponding to an element of G/Γ. If, then, H is the group of permutations of $\theta_1, \theta_2, \dots \theta_l$ which is holoedrically isomorphic with G/Γ, considered as a permutation of rows, every substitution of H must leave $Q(\theta_1, \theta_2, \dots \theta_l)$ unaltered in value. Conversely, if Q is unaltered in value by every substitution of H it must be rational, because in this case every substitution of G leaves it unaltered in value. Therefore (Art. 17) H is the Galoisian group of $a(\theta)=0$; and we may put $H=G/\Gamma$, in the sense that these two groups are holoedrically isomorphic.

Since G/Γ is transitive, H is so too, and hence the equation in θ is irreducible (Art. 19). Moreover, we can prove, as in Art 18, that it is a normal equation, by taking the function

$$\left\{\frac{\phi_1}{\theta-\theta_1} + \frac{\phi_2}{\theta-\theta_2} + \dots + \frac{\phi_l}{\theta-\theta_l}\right\} a(\theta),$$

where ϕ is any rational function of $\theta_1, \theta_2, \dots \theta_l$ and $\phi_1, \phi_2, \dots \phi_l$ are the functions derived from it by applying the substitutions of G/Γ.

24. Consider, now, the effect of adjoining θ_1 to the field of rationality: this means that every function $R(\theta_1; c_1, c_2, \dots c_n)$ which is rational in form is to be considered rational in value. The group Γ is the largest group in G which leaves the values of all such functions unaffected, and it is, in fact, the Galoisian group of $f(x)$ in the new field. To prove this it has to be shown that every rational function $R(x_1, x_2, \dots x_n)$ which has a rational value A in the new field is expressible as an explicit rational function of θ_1.

To prove this, take the function

$$\frac{R(x_1, x_2, \dots x_n)}{\theta - \theta_1},$$

θ being arbitrary, and apply to it all the substitutions of G. Then the function

$$\left\{\frac{R_1}{\theta-\theta_1} + \frac{R_2}{\theta-\theta_2} + \dots + \frac{R_h}{\theta-\theta_h}\right\}(\theta-\theta_1)(\theta-\theta_2)\dots(\theta-\theta_h)$$

can be expressed in a rational form

$$S(\theta; c_1, c_2, \dots c_n)$$

(Arts. 7, 17). Now if

$$a(\theta) = (\theta-\theta_1)(\theta-\theta_2)\dots(\theta-\theta_l)$$

so that $a(\theta) = 0$ is the irreducible rational equation satisfied by θ_1 in the old field, we have

$$(\theta-\theta_1)(\theta-\theta_2)\dots(\theta-\theta_h) = \{a(\theta)\}^k,$$

where $k = h/l$. Moreover, among the denominators

$$(\theta-\theta_1), (\theta-\theta_2), \dots (\theta-\theta_h)$$

only l are distinct, namely,

$$(\theta-\theta_1), (\theta-\theta_2), \dots (\theta-\theta_l).$$

Hence it follows that

$$\left(\frac{R_1}{\theta-\theta_1} + \frac{R_2}{\theta-\theta_2} + \dots + \frac{R_h}{\theta-\theta_h}\right) a(\theta) = T(\theta),$$

where $T(\theta)$ is a rational integral function of θ. If the value of R is unaltered by each substitution of Γ, all the fractions with the denominator $\theta - \theta_1$ must have the same value $\bar{R}$ in the numerator, and we may write, as an arithmetical equality,

$$T(\theta) = \left(\frac{k\bar{R}}{\theta-\theta_1} + \frac{L_2}{\theta-\theta_2} + \frac{L_3}{\theta-\theta_3} + \dots + \frac{L_l}{\theta-\theta_l}\right) a(\theta)$$

true for all values of θ. The quantities $L_2, L_3, \dots L_l$ are all rational functions of $x_1, x_2, \dots x_n$. By putting $\theta = \theta_1$, we obtain

$$\bar{R} = \frac{T(\theta_1)}{ka'(\theta_1)}$$

and this may, if we please, be replaced by an equivalent integral function of degree not exceeding $(l-1)$.

The theorem proved amounts to this :—

If θ is a rational function of the roots of $f(x)=0$, which has for its invariant group a self-conjugate factor, Γ, of G, the effect of adjoining θ to the field of rationality is to reduce the Galoisian group of $f(x)=0$ from G to Γ.

25. In the new field we can construct a new total resolvent for $f(x)$. In fact, if $(v-v_1)$ is any factor of the old resolvent $\psi(v)$, and if the substitutions of Γ give v_1 the values $v_1, v_2, \dots v_k$, then the new total resolvent is

$$F_1(v)=(v-v_1)(v-v_2)\dots(v-v_k)$$
$$=v^k+p_1v^{k-1}+\dots+p_k,$$

where the coefficients are rational in the new field. In one, at least, of these coefficients θ_1 must occur explicitly, because $\psi(v)$ is irreducible in the original field. Moreover

$$\psi(v)=F_1(v)F_2(v)\dots F_l(v),$$

where $F_i(v)$ is obtained from $F_1(v)$ by changing θ_1 to θ_i, then expressing θ_i and its powers in terms of θ_1, and finally reducing the coefficients by means of $a(\theta_1)=0$.

If $F_1(v)$ is reducible in the new field, all its irreducible factors must be of the same degree (cf. Art. 10), and any one of these may be taken as a new primary resolvent. Every root of f may be expressed as a rational function of $v_1, \theta_1, c_1, c_2, \dots c_n, u_1, u_2, \dots u_n$, where v_1 is any root of the new primary resolvent.

26. The equation $a(\theta)=0$ satisfied by the adjoined irrationality θ_1 is usually called a Galoisian resolvent of $f(x)=0$: but we shall find it convenient to call it a Galoisian auxiliary equation, or simply an auxiliary equation when there is no risk of mistake. On the other hand the equation $F_1(v)=0$, obtained in the last article, may be properly called a resolvent.

If we form the auxiliary equation according to the general method of Art. 21, its coefficients will contain the parameters $u_1, u_2, \dots u_n$ in a complicated manner. In any practical case we at once simplify the auxiliary equation as far as we can by giving definite values to the parameters, thus making θ_1 a definite numerical irrationality to be adjoined to the field. It may or may not be convenient to give definite numerical values to the parameters as they occur in $F_1(v)$: for some purposes, even in a practical case, it may be convenient to leave them umbral. This is one of the main reasons for distinguishing between an auxiliary and a resolvent equation: in other respects they are similar, for example they are both normal. The real service rendered by an auxiliary equation is to define a new field of rationality in which the Galoisian group of $f(x)=0$ is of lower order than it was originally, *while at the same time the Galoisian group of the auxiliary equation in the original field is of lower order than that of* $f(x)=0$. Unless this last condition is satisfied, we do not gain anything by the construction of an equation $a(\theta)=0$, even though the adjunction of one of its roots lowers the order of the Galoisian group of f; because in this case the Galoisian group of $a(\theta)=0$ is, in its abstract form, just the same as that of $f(x)=0$, and we are confronted with the original problem in another shape.

If, however, as we have supposed, θ is a rational function of the roots of f which has for its invariant group a proper self-conjugate factor of G (that is, one which is not merely the identical substitution), the problem is really simplified by being made to depend upon two equations

$$a(\theta)=0,$$
$$\psi_1(v, \theta_1)=0,$$

where the first is of order l, a proper factor of h, and has a Galoisian group of order l in the old field; while the second is rational in the field obtained by the adjunction of θ_1, any root of the first, and has a Galoisian group in the new field the order of which is either h/l, or a factor thereof, and is equal in any case to the degree of ψ_1 in v, if we suppose, as we may do, that ψ_1 is irreducible in the new field.

27. As soon as the original Galoisian group of f has been determined, we can construct what is called a *composition-series* for G in the form

$$G, G_1, G_2, \dots G_p, 1,$$

where G_1 is a maximum self-conjugate factor of G, G_2 a maximum self-conjugate factor of G_1, and so on. Using the conventions $G_0=G$, $G_{p+1}=1$, we have a set of indices

$$e_1, e_2, \dots e_p, e_{p+1},$$

such that e_i is the index of G_i with respect to G_{i-1}. The group G_p is simple and its order is e_{p+1}.

We have seen that if we construct a quantity α, which is a rational function of $x_1, x_2, \dots x_n$ and which has G_1 for its invariant group α will satisfy an equation

$$a(\alpha) \equiv \alpha^{e_1} + a_1\alpha^{e_1-1} + \dots + a_{e_1} = 0$$

which is rational and irreducible and normal in the field $(c_1, c_2, \dots c_n)$.

By the adjunction of any one of its roots, we obtain a new field of rationality, which we may denote by (a, c), and in this field the group of f is G_1.

We can now construct a function for which G_2 is the invariant group *in the new field.* Let $t_1, t_2, \dots t_m$ (where $m = h/e_1e_2$) be the elements of G_2, and let θ be an undetermined rational quantity of the new field. We may arrange our notation so that

$$v_1, v_2, \dots v_m$$

are the expressions obtained from v_1 by applying the substitutions of G_2; and then, if we put

$$\beta = (\theta - v_1)(\theta - v_2)\dots(\theta - v_m),$$

β is invariant for G_2 in the field (a, c). By choosing θ properly, as a rational function of a, it will be possible to secure that no other substitution of G_1 leaves β numerically unaltered (cf. Art. 17). Employing a notation which is now usual, we may write

$$G_1 = s_1G_2 + s_2G_2 + \dots + s_{e_2}G_2$$

as an equivalent for a tabular arrangement such as that of Art. 22. Hence we see that the effect of applying all the substitutions of G_1 to β is to produce me_2 expressions which have only e_2 different values, each repeated m times. They are the roots of an equation rational in the new field, and of degree me_2: but since all its roots are of multiplicity m, it is of the form $\{b(\beta)\}^m = 0$, where $b(\beta)$ is also rational, and of degree e_2.

Consequently β is a root of an auxiliary equation

$$b(\beta) \equiv \beta^{e_2} + b_1\beta^{e_2-1} + \dots + b_{e_2} = 0$$

with coefficients which are rational in the new field.

This equation is normal, because G_1/G_2 is a simple and simply transitive group; hence by the adjunction of any one of its roots, all the others become rational, and the Galoisian group of f becomes G_2 in the new field (a, β, c).

Moreover we have

$$F_2(v) = (v - v_1)(v - v_2)\dots(v - v_m) = v^m + q_1v^{m-1} + \dots + q_m$$

a total resolvent for f in the field (a, β, c) with coefficients which are rational in that field. This process may be continued until the Galoisian group of f is reduced to G_p; and finally, by forming an auxiliary equation of degree e_{p+1}, G is reduced to unity, and each root of f is expressible as a rational function of the field $(a, \beta, \dots \lambda, c)$, where $a, \beta, \dots \lambda$ are roots of the $(p+1)$ auxiliary equations. If desirable, this rational function may be transformed so as to be integral in the adjoined irrationalities.

The ith auxiliary equation is of the form

$$x^{e_i} + r_1 x^{e_i - 1} + \dots + r_{e_i} = 0,$$

with coefficients rational in $c_1, c_2, \dots c_n$ and the selected roots of the preceding auxiliary equations.

28. It will be well to illustrate these very important results by a special example. Let the given equation be

$$f(x) \equiv x^6 + x^5 + x^4 + x^3 + x^2 + x + 1 = 0.$$

Then if r is any one of its roots, $r^7 = 1$, and the other roots are r^2, r^3, r^4, r^5, r^6. Thus we have a very simple case of a normal equation. It may be proved that $f(x)$ is irreducible without adjunction: this will, indeed, appear incidentally from what follows.

If we put

$$\begin{aligned}
v_1 &= ar + br^2 + cr^3 + dr^4 + er^5 + fr^6,\\
v_2 &= ar^2 + br^4 + cr^6 + dr + er^3 + fr^5,\\
v_3 &= ar^3 + br^6 + cr^2 + dr^5 + er + fr^4,\\
v_4 &= ar^4 + br + cr^5 + dr^2 + er^6 + fr^3,\\
v_5 &= ar^5 + br^3 + cr + dr^6 + er^4 + fr^2,\\
v_6 &= ar^6 + br^5 + cr^4 + dr^3 + er^2 + fr,
\end{aligned}$$

then v_i is derived from v_1 by changing r to r^i, and $v_1, v_2, \dots v_6$ are the roots of a primary resolvent $\psi(v) = 0$. Expressing the operation of changing v_1 into v_i as a permutation of the roots of f, we have

$$\begin{array}{lll}
s_1 = 1, & s_2 = (124)(365), & s_3 = (132645),\\
s_4 = (142)(356), & s_5 = (154623), & s_6 = (16)(25)(34).
\end{array}$$

These are the elements of the Galoisian group of f, and combine according to the multiplication table

1	s_2	s_3	s_4	s_5	s_6
s_2	s_4	s_6	1	s_3	s_5
s_3	s_6	s_2	s_5	1	s_4
s_4	1	s_5	s_2	s_6	s_3
s_5	s_3	1	s_6	s_4	s_2
s_6	s_5	s_4	s_3	s_2	1

which is to be read $s_2^2 = s_4$, $s_2 s_3 = s_6$, etc. It appears from the table that $s_a s_b = s_b s_a$, so that G is Abelian, and every one of its factors is self-conjugate. As a matter of fact, if we put $s_3 = s$, the elements of G are

$$1,\ s,\ s^2,\ s^3,\ s^4,\ s^5,$$

and the group is cyclical. It is also transitive, so that $f(x)$ is irreducible without adjunction.

One factor of G is $(1, s_2, s_4)$, and from this we can derive an auxiliary quadratic. To find a function of which $(1, s_2, s_4)$ is the invariant group, we start with

$$(t + v_1)(t + v_2)(t + v_4);$$

in this expression the coefficient of $t^2 a$ is

$$r + r^2 + r^4,$$

and this is, in fact, a function such as we require, because (s_3, s_5, s_6) each convert it into

$$r^3 + r^5 + r^6,$$

which has a different value because f is irreducible. If, now, we put

$$y_1 = r + r^2 + r^4, \quad y_2 = r^3 + r^5 + r^6,$$

then $y_1 + y_2 = -1$, and $y_1 y_2 = 2$, in virtue of $f(r) = 0$. Consequently y_1 is a root of the auxiliary equation

$$y^2 + y + 2 = 0 \quad \text{..........................(1).}$$

Let us take
$$y_1 = \frac{-1 + i\sqrt{7}}{2}$$

and adjoin it to the field of rationality, which thus becomes (y_1). The Galoisian group of f reduces to $(1, s_2, s_4)$, of which the only self-conjugate factor is unity. Hence r must be the root of an auxiliary cubic, and since r is changed by s_2, s_4 into r^2, r^4 respectively, this auxiliary cubic is

$$(z - r)(z - r^2)(z - r^4) = 0;$$

or, on multiplying out, and expressing the coefficients in the new field, this is

$$z^3 - y_1 z^2 - (y_1 + 1) z - 1 = 0 \text{....................(2).}$$

If z_1 is any root of this equation, the others are z_1^2, z_1^4; finally the roots of the original equation may be expressed in the form

$$r_1 = z_1, \quad r_2 = z_1^2, \quad r_3 = z_1^3 = y_1 z_1^2 + (y_1 + 1) z_1 + 1,$$
$$r_4 = z_1^4 = -z_1^2 - z_1 + y_1, \quad r_5 = z_1^5 = -(1 + y_1) z_1^2 - z_1 - 1,$$
$$r_6 = z_1^6 = z_1^2 - y_1 z_1 - (1 + y_1).$$

If we solve (2) by the method of Art. 4, we find that

$$(z_1 + \omega z_1^2 + \omega^2 z_1^4)^3$$

is a root of the quadratic

$$t^2 + (2y_1 - 13)\,t - 7\,(2y_1 + 1) = 0,$$

one root of which may be put into the form

$$t = \frac{-2y_1 + 13 + 3\sqrt{21}}{2} = \frac{14 + 3\sqrt{21} - i\sqrt{7}}{2}.$$

Let a definite cube root of this be extracted, and called θ; then since

$$(z_1 + \omega z_1^2 + \omega^2 z_1^4)\,(z_1 + \omega^2 z_1^2 + \omega z_1^4) = 2y_1 + 1 = i\sqrt{7},$$

we may write

$$z_1 + z_1^2 + z_1^4 = \frac{-1 + i\sqrt{7}}{2}, \quad z_1 + \omega^2 z_1 + \omega^2 z_1^4 = \theta,$$

$$z_1 + \omega^2 z_1^2 + \omega z_1^4 = i\sqrt{7}/\theta;$$

whence, by addition,

$$3z_1 = \frac{-1 + i\sqrt{7}}{2} + \theta + \frac{i\sqrt{7}}{\theta}$$

$$= \frac{-1 + i\sqrt{7}}{2} + \theta + \frac{-14 + 3\sqrt{21} + i\sqrt{7}}{14}\,\theta^2.$$

The quantity θ is of the form $\alpha + \beta i$, with α, β real; and the question might be asked, whether α and β admit of representation by means of real radicals. This is not the case, because α is the root of a cubic with all its roots real, so that the formula expressing it again involves cube roots of complex quantities *.

By the adjunction of y_1 the resolvent $\psi(v)$ can be expressed as the product of two rational factors; one of these is

$$F_1(v) \equiv (v - v_1)\,(v - v_2)\,(v - v_4) = v^3 - Pv^2 + Qv - R,$$

where

$$P = (a + b + d)\,y_1 - (c + e + f)\,(1 + y_1),$$

$$\begin{aligned} Q = &-(a^2 + b^2 + d^2)\,(1 + y_1) + (c^2 + e^2 + f^2)\,y_1 \\ &+ (ac + bf + de)\,(2 - y_1) + (ae + bc + df)\,(3 + y_1) \\ &- (bd + da + ab + ef + fc + ce + af + be + cd), \end{aligned}$$

$$\begin{aligned} R = {} & a^3 + b^3 + c^3 + d^3 + e^3 + f^3 \\ &+ (a^2b + a^2c + a^2e + b^2c + b^2d + b^2f + c^2d + c^2e \\ &\qquad + d^2a + d^2e + d^2f + e^2b + e^2f + f^2a + f^2c)\,y_1 \\ &- (a^2d + a^2f + b^2a + b^2e + c^2a + c^2b + c^2f + d^2b + d^2c \\ &\qquad + e^2a + e^2c + e^2d + f^2b + f^2d + f^2e)\,(1 + y_1) \\ &+ (abd + aef + bce + cdf)\,(2 - y_1) \\ &+ (abf + acd + bde + cef)\,(3 + y_1) \\ &- (abc + abe + ace + acf + ade + adf \\ &\qquad + bcd + bcf + bdf + bef + cde + def). \end{aligned}$$

* Hölder, *Mathematische Annalen*, XXXVIII, 307.

The other rational factor may be obtained from this by changing y_1 into $-(1+y_1)$.

This example affords a verification of the theory of Art. 15. The permutations of the parameters which leave $\psi(v)$ *formally* unaltered are

$$\sigma_1 = 1, \qquad \sigma_2 = (dba)(efc), \qquad \sigma_3 = (edfbca),$$
$$\sigma_4 = (bda)(fec), \qquad \sigma_5 = (cbfdea), \qquad \sigma_6 = (fa)(eb)(dc),$$

and these could have been found by experiment from $\psi(v)$, *without assuming any special relations among the roots of* $f(x)$. We should then infer the Galoisian group of $f(x)$ from the permutations σ_i, and hence finally *discover* the relations connecting the roots. The permutations σ which leave $F_1(v)$ unaltered are 1, σ_2, σ_4, as may easily be verified; while σ_3, σ_5, σ_6 each convert $F_1(v)$ into the other rational factor of $\psi(v)$.

Instead of starting with the factor $(1, s_2, s_4)$ we might start with the factor $(1, s_6)$. This leads to the auxiliary equations

$$y^3 + y^2 - 2y - 1 = 0 \quad \text{......................(3)},$$
$$z^2 - y_1 z + 1 = 0 \quad \text{......................(4)},$$

where we may suppose

$$z_1 = r, \quad y_1 = r + r^6.$$

With the notation of Art. 4 we find that $A = B = 7$,

$$\theta^3 = \frac{7 + 21i\sqrt{3}}{2},$$

$$3y_1 = -1 + \theta + 7/\theta = -1 + \theta + \frac{1 - 3i\sqrt{3}}{14}\theta^2,$$

and the reduced forms for the roots are

$$r_1 = z_1, \quad r_2 = z_1^2 = y_1 z_1 - 1, \quad r_3 = z_1^3 = (y_1^2 - 1) z_1 - y_1,$$
$$r_4 = z_1^4 = -(y_1^2 - 1) z_1 - y_1^2 + 1 = -(y_1^2 - 1)(z_1 + 1),$$
$$r_5 = z_1^5 = -y_1 z_1 + y_1^2 - 1, \quad r_6 = z_1^6 = -z_1 + y_1.$$

29. In general, a composition-series for G may be constructed in more ways than one; but in every case the indices $e_1, e_2, \ldots e_{p+1}$ are the same in number and value, and only differ in the order in which they occur*; moreover, the factor-groups G_i/G_{i+1} are the same, except for the order in which they occur, and all of them are simple. Thus the number and the degrees of the auxiliary equations are the same in every case, and however they are formed, the problem of solving them has just the same degree of difficulty. This shows very clearly how deeply the theory of Galois penetrates into the special nature of any given equation.

* Burnside, *Theory of Groups*, pp. 118–123.

A few words may be said as to the effect of adjoining a rational function of the roots, which has for its invariant group Γ, a factor of G which is not self-conjugate. If the order of Γ is k, and we put $h/k = l$, it can be proved, as in Art. 24, that the adjoined function ϕ satisfies a rational equation of degree l, that its Galoisian group is simply isomorphic with the permutations of $(\Gamma, t_2\Gamma, \ldots t_l\Gamma)$ arising from pre-multiplication by substitutions of G, and that the adjunction of ϕ reduces the Galoisian group of f from G to Γ. If we adjoin all the roots of the equation satisfied by ϕ, the group of f sinks to that factor of G which leaves each element of $(\Gamma, t_2\Gamma, \ldots t_l\Gamma)$ unaltered. This factor is the group consisting of all the substitutions common to Γ and its conjugate groups $t_i\Gamma t_i^{-1}$; a group which is self-conjugate in Γ. Consequently, the adjunction of all the roots of the auxiliary equation $a(\phi) = 0$ is equivalent to the adjunction of any rational function for which the self-conjugate group last referred to is the invariant group; hence it is unnecessary to adjoin any irrationalities except those of which the invariant groups are self-conjugate in G.

To avoid misunderstanding, it may be remarked that a group G_i of the composition-series is not necessarily self-conjugate in G; but before constructing the ith auxiliary equation, we have reduced the Galoisian group of f from G to G_{i-1}, and in *this* group G_i is self-conjugate. The advantage of choosing G_i as a *maximum* self-conjugate factor of G_{i-1} is that in this case G_{i-1}/G_i is a simple and simply transitive group *; hence the ith auxiliary equation is normal, and, subject to this condition, of the lowest possible degree.

From what has been said it follows that the natural classification of equations is according to the properties of their Galoisian groups. Equations of quite different degrees are solvable by processes of just the same complexity, provided that their Galoisian groups, in their abstract form, are identical.

30. There is an important theorem which, to a certain extent, forms the converse of that stated in Art. 24, and more generally in Art. 29. It is as follows:—

Suppose that $\phi(y) = 0$ is any rational equation such that the adjunction of one of its roots makes a primary resolvent $\psi(v)$ reducible: then this same reduction may be effected by means of one of the Galoisian auxiliary equations constructed after the manner which has been explained.

* Burnside, pp. 29, 38—40, and Art. 22 above.

We may suppose that $\phi(y)=0$ is irreducible. By hypothesis, $\psi(v)$ becomes reducible in the field (y_1): let the new irreducible factor which has the root v_1 be $\chi(v, y_1)$, a function which must contain y_1 explicity.

With a proper arrangement of the notation, we have identically

$$\chi(v, y_1)=(v-v_1)(v-v_2)\dots(v-v_k).$$

The substitutions $(s_1, s_2, \dots s_k)$ *of* G *which are associated in the usual way with* $v_1, v_2, \dots v_k$ *must form a group* Γ. To see this, we observe that by Art. 7 we may write

$$\chi=\{v-v_1\}\{v-j_2(v_1)\}\dots\{v-j_k(v_1)\},$$

where $j_2, \dots j_k$ denote rational functions. Hence the equation

$$\chi\{j_a(v)\}=0$$

has a root v_1 in common with $\chi(v)=0$, and consequently

$$\chi\{j_a(v_b)\}=0$$

for $b=1, 2, \dots k$. But since s_a and s_b belong to the Galoisian group, we can infer from

$$s_a(v_1)=v_a=j_a(v_1)$$

that

$$s_b(s_a v_1)=j_a(v_b):$$

hence

$$\chi\{s_b(s_a v_1)\}=0$$

and $s_a s_b$ must be one of the set $s_1, s_2, \dots s_k$.

Now let $u(x_1, x_2, \dots x_n)$ be a rational function of the roots of f for which Γ is the invariant group; this will satisfy a rational irreducible equation

$$a(u)=0$$

of degree h/k. We shall have a resolution

$$\psi(v)=\psi_1(v, u)\,\psi_2(v, u)\dots\psi_l(v, u)$$

with $l=h/k$; and we may suppose that $\psi_1(v_1, u)=0$.

Whatever value the rational quantity t may have, the function

$$(t-v_1)(t-v_2)\dots(t-v_k)$$

is invariable for the substitutions of Γ: hence it may be expressed (Art. 24) as an integral function of u and t, say $J(t, u)$. But the function is also $\chi(t, y_1)$: so that the rational equation in u

$$b(u)\equiv J(t, u)-\chi(t, y_1)=0$$

has a root $u=u_1$ in common with $a(u)=0$. By giving t a suitable value we can make u_1 the only common root. The process of finding the highest common factor of $a(u)$ and $b(u)$ leads to an identity

$$Pa+Qb=Ru-S,$$

where R, S are integral functions of y_1; and since a, b have a linear factor in common, we must have

$$Ru_1 - S = 0,$$

$$u_1 = S/R,$$

a rational function of y_1 which may be reduced to an integral form

$$u_1 = \theta(y_1)$$

by means of $$\phi(y_1) = 0.$$

Hence $$\psi_1(v, u_1) = \psi_1\{v, \theta(y_1)\},$$

a rational factor of $\psi(v)$ which vanishes for $v = v_1$, and must therefore coincide with $\chi(v, y_1)$ because χ is irreducible, and the degrees of both factors are the same. This proves that any new irreducible factor of ψ obtained by the adjunction of y_1 can also be obtained by the adjunction of a quantity u_1 which can be expressed as a rational function of the roots of f.

Rational functions of the roots of f have been called by Kronecker *natural* irrationalities (in the case when their values are not rational, of course): thus we may express the theorem by saying that *every possible resolution of the Galoisian resolvent of an equation by means of algebraic operations can be effected by the adjunction of natural irrationalities.*

The roots of a chain of normal Galoisian auxiliary equations are natural irrationalities: in a certain sense they form a "simplest" set of irrationalities in terms of which all the roots of the given equation can be rationally expressed.

CHAPTER II

CYCLICAL EQUATIONS

31. THE only irreducible equations which have unity for their Galoisian group are linear, and require no discussion. The next simplest irreducible equation is one of which the Galoisian group is cyclical, so that

$$G = (1, s, s^2, \dots s^{n-1})$$

with $$s^n = 1.$$

This is called a cyclical equation. The necessary and sufficient condition that a rational function of its roots should have a rational value is that its value remains unaltered when the substitution s is applied to it.

The group G must be transitive, since f is supposed to be irreducible: hence s must consist of a single cycle which, with a suitable notation for the roots, may be written in the forms

$$s = (x_1 x_2 \dots x_n) = (12 \dots n).$$

If p is any prime factor of n, and $n = mp$, the group

$$G_1 = 1, s^p, s^{2p}, \dots s^{(m-1)p}$$

is self-conjugate in G, and we can form an auxiliary equation

$$a(\alpha) = 0,$$

of degree p, which reduces the group of f to G_1.

If q is any prime factor of m, and $m = lq$, the group

$$G_2 = 1, s^{pq}, s^{2pq}, \dots s^{(l-1)pq}$$

is self-conjugate in G_1, and we can form another auxiliary equation

$$b(\beta) = 0$$

of degree q, with coefficients rational in the field (α), which reduces the group of f to G_2: and so on.

It thus appears that if

$$n = p^h q^k \dots z^t,$$

where $p, q, \dots z$ are different primes, the complete solution of $f = 0$ can be obtained from $(h + k + \dots + t)$ auxiliary equations: h of these are of degree p, k of degree q, ... t of degree z.

Each of the auxiliary equations is cyclical. For example, the group of $b(\beta)$ is G_1/G_2, and this is cyclical, because if we break up G_1 into parts (or rows) with respect to G_2 we have

$$G_1 = G_2 + s^p G_2 + s^{2p} G_2 + \dots + s^{(q-1)p} G_2,$$

and hence $$s^{ip} G_1 = s^{ip} G_2 + s^{(i+1)p} G_2 + \dots + s^{(q-1+i)p} G_2$$

a cyclical permutation of the parts. In other words, the group of b is of the form $(1, \sigma, \sigma^2, \dots \sigma^{q-1})$ with $\sigma^q = 1$, and so for any other auxiliary.

32. Thus the solution of any cyclical equation may be made to depend upon the solution of auxiliary cyclical equations of prime degrees. In the first place, however, we shall explain a process of solution which is applicable to the original equation as well as to its auxiliaries. This solution expresses the roots of f rationally in terms of its coefficients, a primitive nth root of unity ϵ, and the nth root of a quantity which is rational when ϵ is adjoined to the original field.

Let $$\theta_1 = x_1 + \epsilon x_2 + \dots + \epsilon^{n-1} x_n :$$

then $$s\theta_1 = x_2 + \epsilon x_3 + \dots + \epsilon^{n-2} x_n + \epsilon^{n-1} x_1 = \epsilon^{-1}\theta_1 :$$

and similarly $$s^i\theta_1 = \epsilon^{-i}\theta_1 :$$

hence $$s^i(\theta_1^n) = \epsilon^{-in}\theta_1^n = \theta_1^n,$$

and θ_1^n must be a rational quantity in the new field, because its value is unaffected by any substitution of G, and the group of f in the new field must be either G itself, or a factor thereof. Consequently we may put

$$\theta_1 = \sqrt[n]{R},$$

where $\sqrt[n]{R}$ denotes some one definite nth root of the rational quantity R, for instance the real root, if it exist. R may, and in general will, explicitly contain the auxiliary quantity ϵ.

Now consider the expression obtained from θ_1 by changing ϵ to ϵ^k where k is any positive integer. Calling it θ_k, we have

$$\theta_k = x_1 + \epsilon^k x_2 + \dots + \epsilon^{(n-1)k} x_n,$$

$$s\theta_k = \epsilon^{-k}\theta_k,$$

and hence $$s\left(\frac{\theta_k}{\theta_1^k}\right) = \frac{\epsilon^{-k}\theta_k}{\epsilon^{-k}\theta_1^k} = \frac{\theta_k}{\theta_1^k}.$$

Assuming that the value of θ_1 is not zero, it follows that

$$\theta_k = B_k \theta_1^k, \qquad (k = 2, 3, \dots \overline{n-1})$$

where B_k is a rational quantity in the new field.

Finally
$$nx_1 = -c_1 + \theta_1 + \theta_2 + \dots + \theta_{n-1}$$
$$= -c_1 + \theta_1 + B_2\theta_1^2 + B_3\theta_1^3 + \dots + B_{n-1}\theta_1^{n-1}.$$

By changing θ_1 into $\epsilon^{-i}\theta_1$ we obtain a similar expression for x_{i+1}.

As an illustration, take the example of Art. 28. In the first mode of solution, after the adjunction of y_1,

$$\theta_1^3 = \frac{14 + 3\sqrt{21} - i\sqrt{7}}{2} = 5 - 4y_1 - (3 + 6y_1)\,\omega,$$

$$3z_1 = y_1 + \theta_1 - \frac{8 + 2y_1 + (3 + 6y_1)\,\omega}{7}\,\theta_1^2,$$

$$3z_2 = y_1 + \omega^2\theta_1 + \frac{3 + 6y_1 - (5 - 4y_1)\,\omega}{7}\,\theta_1^2,$$

$$3z_3 = y_1 + \omega\theta_1 + \frac{5 - 4y_1 + (8 + 2y_1)\,\omega}{7}\,\theta_1^2.$$

In the second mode of solution

$$\theta_1^3 = \frac{7 + 21i\sqrt{3}}{2} = 7\,(3\omega + 2),$$

and the roots of the first auxiliary equation are given by

$$3y_1 = -1 + \theta_1 - \frac{1 + 3\omega}{7}\,\theta_1^2,$$

$$3y_2 = -1 + \omega^2\theta_1 + \frac{3 + 2\omega}{7}\,\theta_1^2,$$

$$3y_3 = -1 + \omega\theta_1 - \frac{2 - \omega}{7}\,\theta_1^2.$$

33. The method above explained breaks down when $\theta_1 = 0$ for each primitive root ϵ. To avoid this difficulty, Weber* has put the expression for x_1 into a slightly different form as follows.

We have identically

$$nx_1 + c_1 = \Sigma\theta_i, \qquad (i = 1, 2, \dots \overline{n-1})$$
$$nx_h + c_1 = \Sigma\epsilon^{-hi}\theta_i,$$

and hence
$$n\,(x_h - x_1) = \Sigma\,(\epsilon^{-hi} - 1)\,\theta_i.$$

Now let $h = n/p$, where p is any prime factor of n; the coefficient $(\epsilon^{-hi} - 1)$ vanishes whenever i is a multiple of p, while on the other

* *Algebra*, I, 589.

hand $(x_h - x_1)$ is not zero, because f is irreducible. Consequently there must be one integer i at least such that θ_i does not vanish, and i is prime to p.

If therefore
$$n = p^\alpha q^\beta r^\gamma \ldots,$$
where $p, q, r, \ldots$ are different primes, we can find integers $\lambda, \mu, \nu, \ldots$ prime to $p, q, r, \ldots$ respectively, such that $\theta_\lambda, \theta_\mu, \theta_\nu$, etc. are all different from zero.

Taking any positive integers $t, x, y, z, \ldots$ and denoting, as before, the generating substitution of G by s, we have
$$s(\theta_t \theta_\lambda^{-x} \theta_\mu^{-y} \theta_\nu^{-z} \ldots) = \epsilon^u \theta_t \theta_\lambda^{-x} \theta_\mu^{-y} \theta_\nu^{-z} \ldots,$$
where
$$u = -t + \lambda x + \mu y + \nu z + \ldots.$$

The greatest common measure of λ, μ, ν, etc. is prime to n: consequently there are positive integers ξ, η, ζ, etc. such that
$$\lambda\xi + \mu\eta + \nu\zeta + \ldots \equiv 1; \qquad \text{(mod } n)$$
and if we put
$$\theta = \theta_\lambda^\xi \theta_\mu^\eta \theta_\nu^\zeta \ldots,$$
θ is a quantity *which does not vanish* and is such that
$$s(\theta_t \theta^{-t}) = \theta_t \theta^{-t}.$$

Consequently
$$\theta_t = R_t \theta^t \qquad (t = 1, 2, \ldots \overline{n-1})$$
where R_t is rational; and
$$n x_i = -c_1 + \Sigma \epsilon^{-u} \theta_t$$
$$= -c_1 + \Sigma \epsilon^{-u} R_t \theta^t$$
with
$$\theta^n = R,$$
where R is a non-vanishing quantity, rational in the new field.

34. Since $s(\theta_\lambda^x) = \epsilon^{-\lambda x} \theta_\lambda^x$, the lowest power of θ_λ which is rational is determined by the congruence
$$\lambda x \equiv 0 \qquad \text{(mod } n)$$
or
$$x \equiv 0 \qquad \text{(mod } n/d)$$
where $d = \text{dv}(n, \lambda)$. On account of λ being prime to p, d is also prime to p, and we may write
$$n/d = p^\alpha l_1$$
where l_1 is an integer. If we put
$$\theta_\lambda^{l_1} - \phi_\lambda,$$
then
$$\phi_\lambda^{p^\alpha} = T_\lambda,$$

where T_λ is rational, and

$$s(\phi_\lambda) = \epsilon^{-l_1\lambda}\phi_\lambda.$$

We may in the same way derive from θ_μ, θ_ν, etc. quantities ϕ_μ, ϕ_ν, etc. such that

$$\phi_\mu^{q^\beta} = T_\mu, \quad \phi_\nu^{r^\gamma} = T_\nu, \dots,$$

$$s(\phi_\mu) = \epsilon^{-m_1\mu}\phi_\mu, \quad s(\phi_\nu) = \epsilon^{-n_1\nu}\phi_\nu, \dots$$

and so on.

The integer l_1 is prime to p, m_1 is prime to q, and so on: hence we can find integers ξ, η, ζ, etc. such that

$$l_1\lambda\xi + m_1\mu\eta + n_1\nu\zeta + \dots \equiv 1. \qquad \text{(mod } n\text{)}$$

Now $$s(\theta_t\phi_\lambda^{-x}\phi_\mu^{-y}\phi_\nu^{-z}\dots) = \epsilon^u\theta_t\phi_\lambda^{-x}\phi_\mu^{-y}\phi_\nu^{-z},$$

where $$u = -t + l_1\lambda x + m_1\mu y + n_1\nu z + \dots;$$

so that $u \equiv 0$ (mod n) if

$$x, y, z, \dots = t\xi, t\eta, t\zeta, \dots.$$

Consequently, if we put

$$\phi = \phi_\lambda^\xi\phi_\mu^\eta\phi_\nu^\zeta\dots,$$

then $$\theta_t = S_t\phi^t$$

where S_t is rational:

$$nx_1 + c_1 = S_1\phi + S_2\phi^2 + \dots + S_n\phi^{n-1} \quad \dots\dots\dots\dots\dots(1),$$

and ϕ , ϕ_μ, etc. are determined by the binomial equations

$$\phi_\lambda^{p^\alpha} = T_\lambda, \quad \phi_\mu^{q^\beta} = T_\mu, \dots,$$

the degrees of which are the powers of primes which occur in n. By giving ϕ_λ, ϕ_μ, etc. all their different values, ϕ assumes n different values, and if these are substituted in (1), we get all the roots of the given equation. Of course the adjunction of the quantities ϕ_λ, ϕ_μ, etc. is equivalent to the adjunction of the single quantity θ which is determined by a binomial equation of degree n; but the equations which determine ϕ_λ, etc. are all lower than the one which determines θ. *In this respect* the last form of the solution may be considered the simpler one. All this illustrates the fact that what is to be called the "simplest" solution of an equation is partly a matter of convention.

Thus, again, if, in the present case, we solve the equation by a chain of Galoisian auxiliaries, they will all be of prime degree, and for each of them one at least of the quantities θ_t must be different from zero, so that Weber's supplementary transformation is unnecessary. In these respects the solution is the simplest of all: on the other hand, just because the expressions for the roots are more explicit, they are more complicated in appearance.

35. In the solution of the general cyclic equation complex roots of unity appear as auxiliary irrationalities. These roots of unity are themselves the roots of cyclic (or Abelian) equations, and it is natural to inquire how far the solution of these special equations can be carried.

If
$$n = p_1 p_2 \dots p_h,$$
where p_1, p_2, etc. are powers of different primes, the complex roots of $x^n = 1$ may all be expressed in the form
$$\alpha^x \beta^y \dots \lambda^z,$$
where $\alpha, \beta, \dots \lambda$ are roots of
$$x^{p_1} = 1, \quad x^{p_2} = 1, \dots x^{p_h} = 1,$$
so that it is sufficient to consider the case in which n is a power of a prime.

We shall begin by supposing that $n = p$, an odd prime; the equation to be solved is therefore
$$f(x) = x^{p-1} + x^{p-2} + \dots + x + 1 = 0.$$

If r is any one of its roots, the others are $r^2, r^3, \dots r^{p-1}$. These may be expressed in a more convenient form as follows. Let g be a primitive root of p; that is to say, a primitive root of the congruence
$$g^{p-1} \equiv 1. \qquad \pmod p$$

Then $1, g, g^2, \dots g^{p-2}$ form a complete set of residues of p, and if we write
$$r_i = r^{g^{i-1}}$$
the roots of $f(x)$ will be denoted by suffixes in such a way that
$$r_{i+1} = r_i^g.$$

In this notation, every integral function of the roots which is unaltered in value by the substitution
$$s = (r_1 r_2 \dots r_{p-1})$$
is rational.

The function in question can be reduced to the form $\phi(r_1)$, where ϕ is a rational polynomial. If the substitution s is applied to the original form of the function, its effect is the same as changing r_1 into r_1^g in $\phi(r_1)$. Hence if A is the value of the function, which by hypothesis is unaltered,
$$\begin{aligned} A &= \phi(r_1) = \phi(r_1^g) = \phi(r_1^{g^2}) = \dots \\ &= \phi(r_1) = \phi(r_2) = \dots \\ &= \frac{1}{p-1}\{\phi(r_1) + \phi(r_2) + \dots + \phi(r_{p-1})\}, \end{aligned}$$
a rational quantity, because symmetrical in the roots of f.

If $(v - v_1)$ is a factor of the total resolvent of f and we put $s^h(v_1) = v_{h+1}$, the factor

$$\psi(v) = (v - v_1)(v - v_2) \dots (v - v_{p-1})$$

will be rational, and moreover it will be irreducible, because otherwise there would be an identity

$$v^h - v^{h-1}(au_1 + \dots) + \dots = (v - v_1)(v - v_a)(v - v_\beta) \dots$$
$$= v^h - v^{h-1}\{(r_1 + r_{a+1} + r_{\beta+1} + \dots)u_1 + \dots\} + \dots$$

leading to $\qquad r_1 + r_{a+1} + r_{\beta+1} + \dots = a$

with a rational, and less than $(p-1)$ terms on the left-hand side. This is impossible, because $f(x) = 0$ is an irreducible equation*.

Hence $\psi(v)$ is a primary resolvent of f, and the Galoisian group of f is $(1, s, s^2, \dots s^{p-2})$, so that f is a cyclical equation. We may proceed to solve it, either by forming a chain of auxiliary equations, the degrees of which are the prime factors of $(p-1)$, or else by adjoining a primitive $(p-1)$th root of unity, and proceeding as in Arts. 32, 33.

36. An example of the first method (for $p = 7$) has been completely worked out in Art. 28. In the general case, let $p - 1 = ef$, where e is a prime. Putting

$$\alpha = r_1 + r_{e+1} + r_{2e+1} + \dots + r_{p-e},$$

α will be a root of an auxiliary equation

$$a(\alpha) = 0,$$

with rational integral coefficients and of degree e

If $f = gh$, where g is a prime, we put

$$\beta = r_1 + r_{ge+1} + r_{2ge+1} + \dots + r_{p-ge},$$

and now β is a root of an auxiliary equation

$$b(\beta) = 0$$

of degree g, with coefficients which are rational polynomials in α. We proceed in this way until all the prime factors of $(p-1)$ are exhausted.

A case of historical interest is when $p = 17$. The auxiliary equations are (taking 3 as the primitive root of 17)

$$\alpha^2 + \alpha - 4 = 0,$$

$$\beta^2 - \alpha\beta - 1 = 0,$$

$$2\gamma^2 - 2\beta\gamma + (\alpha\beta - \alpha + \beta - 3) = 0$$

$$\delta^2 - \gamma\delta + 1 = 0.$$

* Weber, *Algebra*, I, 596; or my *Theory of Numbers*, p. 186.

All these equations, except the last, have real roots, and $\alpha, \beta, \gamma, \delta$ can all be obtained explicitly in forms containing real arithmetical surds: thus we may put

$$\alpha = \frac{-1+\sqrt{17}}{2}, \quad \beta = \frac{-1+\sqrt{17}+\sqrt{(34-2\sqrt{17})}}{4},$$

but the expressions for γ and δ are too complicated to be worth writing down.

37. To solve the equation considered in Art. 28 by the method of Art. 32, we put

$$\theta_1 = r + \omega^2 r^2 - \omega r^3 + \omega r^4 - \omega^2 r^5 - r^6$$

($-\omega$ being a primitive sixth root of unity, and the cyclical order of the roots of f being $r, r^3, r^2, r^6, r^4, r^5$ when we take 3 as the primitive root of 7). It is found by actual multiplication that

$$\theta_1^3 = (5-3\omega)(r + r^2 + r^4 - r^3 - r^5 - r^6),$$

$$\theta_1^6 = -7(16-39\omega) = (1+3\omega)(2+3\omega)^5,$$

where it may be observed that in the field (ω) the norm of θ_1^6 is 7^6. It may also be verified that

$$\theta_2 = \frac{2-\omega}{7}\theta_1^2, \quad \theta_3 = \frac{8+3\omega}{49}\theta_1^3, \quad \theta_4 = -\frac{18+19\omega}{343}\theta_1^4,$$

$$\theta_5 = \frac{\theta_2\theta_3}{1-2\omega} = \frac{55+39\omega}{2401}\theta_1^5;$$

so that finally

$$r = -\frac{1}{7} + \frac{\theta_1}{7} + \frac{2-\omega}{7^2}\theta_1^2 + \frac{8+3\omega}{7^3}\theta_1^3 - \frac{18+19\omega}{7^4}\theta_1^4 + \frac{55+39\omega}{7^5}\theta_1^5$$

38. The simplest way of calculating the quantities θ_i is the following. If h is any one of the numbers 1, 2, 3, ... $(p-3)$, the product $\theta_1\theta_h$ is not rational, and its quotient by θ_{h+1} is equal to the coefficient of r in the product $\theta_1\theta_h$ after reducing it by *first* replacing all powers of r higher than r^{p-1} according to the formula $r^{ap+b} = r^b$, and *then* replacing any rational term a by its equivalent value

$$-a(r + r^2 + \ldots + r^{p-1}).$$

Now $$\theta_1\theta_h = \Sigma \epsilon^{\text{ind}\, a + h\, \text{ind}\, b} r^{a+b}. \quad (a, b = 1, 2, \ldots \overline{p-1})$$

The only pairs (a, b) which contribute to the coefficient which we wish to find are those for which

$$a + b = p,$$

$$a + b = p + 1$$

The second set contributes, after the first reduction, a coefficient of r which is

$$\Sigma\epsilon^{\text{ind}\, a+h\, \text{ind}\,(p+1-a)}, \qquad (a=2, \dots \overline{p-1})$$

the other set, after the second reduction, contributes

$$-\Sigma\epsilon^{\text{ind}\, a+h\, \text{ind}\,(p-a)}. \qquad (a=1, 2, \dots \overline{p-1})$$

Since
$$\text{ind}\,(p-a) = \text{ind}\,(-a)$$
$$= \tfrac{1}{2}(p-1) + \text{ind}\, a,$$

the sum last written

$$= \epsilon^{\frac{1}{2}h(p-1)} \sum_{a=1}^{a=p-1} \epsilon^{(h+1)\,\text{ind}\, a} = 0;$$

and hence
$$\frac{\theta_1\theta_h}{\theta_{h+1}} = \Sigma\epsilon^{\text{ind}\, a+h\, \text{ind}\,(p+1-a)} \qquad (a=2, 3, \dots \overline{p-1})$$

On the other hand, if $h=p-2$, then $\theta_1\theta_h$ is rational. Its value may be written in the form

$$\theta_1\theta_{p-2} = \Sigma\epsilon^{\text{ind}\, a+(p-2)\,\text{ind}\, b}\, r^{a+b}$$
$$= \Sigma\epsilon^{\text{ind}\, a-\text{ind}\, b}\, r^{a+b}$$

since $\epsilon^{p-1}=1$. Now if we put $a \equiv tb \pmod p$, we obtain the equivalent expression

$$\Sigma\epsilon^{\text{ind}\, t}\, r^{(1+t)b} \qquad (b, t = 1, 2, \dots \overline{p-1})$$

The terms for which $t=p-1$ contribute

$$(p-1)\,\epsilon^{\text{ind}\,(p-1)} = -p+1;$$

for any other value of t

$$\sum_b r^{(1+t)b} = \frac{r^{1+t}\{r^{(1+t)(p-1)}-1\}}{r^{1+t}-1} = -1,$$

hence the value of all the remaining terms

$$= -1 \sum_{t=1}^{t=p-2} \epsilon^{\text{ind}\, t} = -1;$$

and finally
$$\theta_1\theta_{p-2} = -p.$$

This, together with

$$\frac{\theta_1\theta_h}{\theta_{h+1}} = \sum_{m=2}^{m=p-1} \epsilon^{\text{ind}\, m+h\, \text{ind}\,(p+1-m)}, \qquad [h=1, 2, \dots \overline{p-3}]$$

enables us to find the values of $\theta_1, \theta_2, \dots \theta_{p-2}$ with great facility. Of course the indices of the powers of ϵ are reduced, at the first opportunity, to their least residues, mod $(p-1)$.

As an example, when $p=7$, we construct the table of indices for the primitive root 3 :—

m	1	2	3	4	5	6
ind m	6	2	1	4	5	3
ind $(8-m)$		3	5	4	1	2

and hence find

$$\frac{\theta_1^2}{\theta_2}=\epsilon^5+1+\epsilon^2+1+\epsilon^5=3+\omega,$$

$$\frac{\theta_1\theta_2}{\theta_3}=\epsilon^2+\epsilon^5+1+\epsilon+\epsilon=1-2\omega,$$

$$\frac{\theta_1\theta_3}{\theta_4}=\epsilon^5+\epsilon^4+\epsilon^4+\epsilon^2+\epsilon^3=-1+2\omega,$$

$$\frac{\theta_1\theta_4}{\theta_5}=\epsilon^3+\epsilon^3+\epsilon^2+\epsilon^3+\epsilon^5=-3-\omega,$$

$$\theta_1\theta_5=-7.$$

By multiplication we find that

$$\theta_1^6=-7(1-2\omega)^2(3+\omega)^2$$
$$=-7(16-39\omega)$$

as before; and all the results of Art. 37 may now be obtained with ease.

39. Suppose now that $n=p^a$, a power of a prime. The primitive nth roots of unity in this case are the roots of the equation

$$f(x)=\frac{x^{p^a}-1}{x^{p^{a-1}}-1}=x^{p^{a-1}(p-1)}+x^{p^{a-1}(p-2)}+\ldots+x^{p^{a-1}}+1=0,$$

which is irreducible, and of degree $p^{a-1}(p-1)$. It is also cyclical, because there are primitive roots of p^a which can be used, as in the case when $a=1$, to fix a cyclical order of the roots, and the arguments of Art. 35 may be repeated. The indices of the composition-series will be the prime factors of $(p-1)$ and also the prime p repeated $(a-1)$ times. Hence if we solve the equation $f(x)=0$ by a chain of Galoisian auxiliaries, $(a-1)$ of these will be of degree p, and (Art. 30) no purely algebraical solution can replace these auxiliaries by others of lower degree.

Finally, if

$$n = p^{\alpha} q^{\beta} r^{\gamma} \ldots, \qquad \phi(n) = p^{\alpha-1}(p-1)\, q^{\beta-1}(q-1) \ldots,$$

the primitive nth roots of unity are $\phi(n)$ in number, and they may be determined by as many chains of auxiliary equations as there are different prime factors of n. The degrees of the auxiliary equations are the prime factors of $\phi(n)$. It should be observed that the primitive nth roots satisfy an irreducible equation of degree $\phi(n)$, but this equation is not cyclical.

A specially interesting case is when the auxiliary equations are all quadratics. The necessary and sufficient condition for this is that $\phi(n)$ should be a power of 2; this is equivalent to saying that

$$n = 2^{\alpha} pqr \ldots,$$

where p, q, r, etc. are different primes, each of the form $2^m + 1$. When n is of this form, and then only, a regular polygon of n sides can be inscribed in a circle by means of the rule and compass; because the complete solution of $x^n = 1$ leads to the determination of $\cos 2\pi/n$ and $\sin 2\pi/n$, and conversely, while every construction with rule and compass can be put into an analytical form which involves only linear and quadratic equations. This remarkable connexion between geometry and analysis was discovered by Gauss.

The values of n, below 100, which are of this special form are

3, 4, 5, 6, 8, 10, 12, 15, 16, 17, 20, 24, 30,
32, 34, 40, 48, 51, 60, 64, 68, 80, 85, 96.

Of these the only ones which are not considered in Euclid's *Elements*, or at least easily brought into connexion with the cases ($n = 3$, 4, 5, 6, 15) which he does consider, are 17, 34, 51, 68 and 85.

CHAPTER III

ABELIAN EQUATIONS

40. A GROUP is said to be *Abelian* when its elements satisfy the commutative law of multiplication: that is to say when $ss' = s's$, s and s' denoting any two elements of the group. An Abelian equation is one of which the Galoisian group is Abelian. Cyclical equations form the simplest class of Abelian equations: it will be shown in this chapter that every Abelian equation may be solved by means of auxiliary cyclical equations.

It will be supposed, in the first place, that the given Abelian equation is irreducible. This being so, its Galoisian group G is transitive, and will contain a substitution s_i which converts x_1 into any other assigned root x_i.

The substitutions of G which leave x_1 unaltered form a sub-group of G. Let σ be any one of these: then since s_i^{-1} changes x_i into x_1,

$$s_i^{-1}\sigma s_i(x_i) = \sigma s_i(x_1) = s_i\{\sigma(x_1)\} = x_i,$$

that is to say, $s_i^{-1}\sigma s_i$ leaves x_i unaltered. But since G is Abelian, $s_i^{-1}\sigma s_i = s_i^{-1}s_i\sigma = \sigma$; consequently σ leaves every root unaltered, and is the identical substitution. It follows from this that G is simply transitive, and that if $x_1, x_2, \ldots x_n$ are the roots of the given equation

$$G = (1, s_2, s_3, \ldots s_n)$$

where s_i is the definite substitution which changes x_1 into x_i.

Moreover the adjunction of x_1 reduces G to unity: consequently $x_2, \ldots x_n$ are expressible as rational functions of x_1, and $f(x) = 0$ is a normal equation.

Let the rational expressions of the other roots in terms of x_1 be

$$x_2 = \theta_2(x_1),\quad x_3 = \theta_3(x_1),\quad \ldots\quad x_n = \theta_n(x_1).$$

To these equations (Art. 17) we may apply any substitution of G: thus from

$$x_i = \theta_i(x_1),\quad x_j = \theta_j(x_1)$$

we deduce

$$s_j x_i = \theta_i(x_j),\quad s_i x_j = \theta_j(x_i).$$

But $$s_j x_i = s_j \{s_i x_1\} = s_i \{s_j x_1\} = s_i x_j:$$
consequently $$\theta_i(x_j) = \theta_j(x_i),$$
that is, $$\theta_i\{\theta_j(x_1)\} = \theta_j\{\theta_i(x_1)\}.$$

By applying a Galoisian substitution to this we infer that

$$\theta_i\{\theta_j(x_k)\} = \theta_j\{\theta_i(x_k)\} \qquad [i, j, k = 1, 2, \ldots n]$$

with the convention that $\theta_1(x_k) = x_k$.

In other words, the rational function

$$\theta_i\{\theta_j(x)\} - \theta_j\{\theta_i(x)\}$$

must either vanish identically, or have a numerator which is divisible by $f(x)$. In general, it is the latter case that occurs; so we may write, to express this fact,

$$\theta_i\{\theta_j(x)\} - \theta_j\{\theta_i(x)\} \equiv 0. \qquad [\text{mod} f(x)]$$

Conversely if the roots of a normal equation $f(x) = 0$ can be expressed in a form $x_i = \theta_i(x_1)$ such that these congruences are satisfied, the Galoisian group is Abelian. For we have arithmetically

$$\theta_i\{\theta_j(x_1)\} = \theta_j\{\theta_i(x_1)\}:$$
that is $$\theta_i(x_j) = \theta_j(x_i):$$
but since $$s_i x_1 = x_i = \theta_i(x_1), \text{ and } s_j x_1 = x_j = \theta_j(x_1),$$
it follows that $$s_j(s_i x_1) = \theta_i(x_j), \quad s_i(s_j x_1) = \theta_j(x_i);$$
consequently $$s_j(s_i x_1) = s_i(s_j x_1),$$
and in this we may change x_1 to x_k. Finally, then, $s_i s_j = s_j s_i$ identically, and the group of the equation is Abelian. It will be observed that this converse theorem is true whether $f(x)$ is irreducible or not.

41. The simplest way of expressing the elements of an Abelian group is by what is called a basis*. The elements $s_1, s_2, \ldots s_h$ form a basis of G when every element of G can be expressed in one and only one way in the form

$$s_1^x s_2^y \ldots s_h^t \qquad (x \leqslant m_1, y \leqslant m_2, \ldots t \leqslant m_h)$$

with $x, y, \ldots t$ positive integers, and $m_1, m_2, \ldots m_h$ the least positive integers such that

$$s_1^{m_1} = s_2^{m_2} = \ldots = s_h^{m_h} = 1.$$

If desirable, the base may be so chosen that $m_1, m_2, \ldots m_h$ are powers of primes: of course their product is equal to n, the order of G.

* Weber, *Algebra*, II, 38–45.

42. No generality will be lost, and the notation will be much simplified, if we suppose that the basis of G consists of three elements s, t, u, of order a, b, c respectively, so that $abc = n$, and all the elements of G are expressed by

$$s^i t^j u^k. \qquad (i \leqslant a,\ j \leqslant b,\ k \leqslant c)$$

Let p be any prime factor of a; then the substitutions for which, in their basic form, i is divisible by p form a self-conjugate sub-group of G, the index of which, with respect to G, is p. Since p is prime, this is a maximum sub-group, which we may denote by G_1, and a rational function of the roots for which G_1 is the invariant group will satisfy a rational *cyclic* equation of degree p. By adjoining one root of this equation, the Galoisian group of f sinks from G to G_1.

Suppose, now, that q is a prime factor of a/p: then the substitutions of G which, in their basic form, are such that i is divisible by pq, form a maximum self-conjugate factor of G_1, which we may call G_2. A function for which G_2 is the invariant group in the enlarged field will satisfy a rational cyclical equation of order q, and the adjunction of one of its roots reduces the group of f from G_1 to G_2. By proceeding in this way, we can exhaust all the prime factors of a and reduce the group of f to those substitutions of which the basic forms are $t^j u^k$. If p' is any prime factor of b we have a group $(t^j u^k)$ with j divisible by p', and a corresponding cyclic auxiliary of degree p', and so on. The group of f is finally reduced to unity by a chain of auxiliary cyclic equations, the degrees of which are the prime factors of n: that is to say, if $n = p^\alpha q^\beta r^\gamma \ldots$, there will be α auxiliary equations of degree p, β of degree q, γ of degree r, etc.

43. As a simple illustration, we will take

$$f(x) = x^8 - x^6 + x^4 - x^2 + 1 = 0$$

the roots of which are the primitive 20th roots of unity. If we arrange the roots so that

$$x_1 = r, \quad x_2 = r^3, \quad x_3 = r^7, \quad x_4 = r^9,$$
$$x_5 = r^{11}, \quad x_6 = r^{13}, \quad x_7 = r^{17}, \quad x_8 = r^{19},$$

the substitutions of G are

$s_1 = 1,$

$s_2 = (1243)(5687), \qquad s_3 = (1342)(5786),$

$s_4 = (14)(23)(58)(67), \qquad s_5 = (15)(26)(37)(48),$

$s_6 = (1647)(2835), \qquad s_7 = (1746)(2538),$

$s_8 = (18)(27)(36)(45).$

If we apply these to the function given by

$$y_1 = x_1 + x_2 + x_3 + x_4$$

the only new function arising is

$$y_2 = x_5 + x_6 + x_7 + x_8 = -y_1.$$

Hence y_1 is a root of a rational quadratic. To find it, we have, with the help of $f(r) = 0$,

$$y_1 = r + r^3 + r^7 + r^9 = 2r^7 - r^5 + 2r^3,$$

$$y_1^2 = (2r^7 - r^5 + 2r^3)^2 \equiv 5r^{10} \quad [\text{mod } f(r)]$$

$$= -5,$$

and the first auxiliary equation is

$$y_1^2 + 5 = 0.$$

If we now put

$$z_1 = x_1 + x_4 = r + r^9, \quad z_2 = x_2 + x_3 = r^3 + r^7$$

we find that $z_1 + z_2 = y_1$, $z_1 z_2 = -1$, so that the second auxiliary equation is

$$z_1^2 - y_1 z_1 - 1 = 0.$$

Finally x_1 and x_4 are the roots of

$$x_1^2 - z_1 x_1 - 1 = 0.$$

By actually solving the auxiliaries we see that we may take

$$y_1 = i\sqrt{5}, \quad z_1 = \frac{i(\sqrt{5}+1)}{2}, \quad x_1 = \frac{\sqrt{(10-2\sqrt{5})} + i(\sqrt{5}+1)}{4};$$

and as a verification we observe that the expression last written is $\exp(6\pi i/20)$, one of the primitive roots required.

The group G is in this case dibasic: if we put

$$s = s_2, \quad t = s_8,$$

then (s, t) is a basis, and the basic representation of G is

$$s_1 = 1, \quad s_2 = s, \quad s_3 = s^3, \quad s_4 = s^2,$$

$$s_5 = s^2 t, \quad s_6 = s^3 t, \quad s_7 = st, \quad s_8 = t,$$

with

$$s^4 = t^2 = 1.$$

It is a very remarkable fact, discovered by Kronecker, that if the coefficients of an Abelian equation are ordinary real integers, its roots can be expressed as rational functions of roots of unity, with real rational coefficients. Proofs of this theorem have been given by Weber and Hilbert, but they are too long and difficult to be reproduced here.

CHAPTER IV

METACYCLIC EQUATIONS

QUINTIC AND SEXTIC EQUATIONS

44. SUPPOSE that p is a prime number, and that g is any one of its primitive roots. The numbers $(1, 2, 3, \ldots p)$ form a complete system of residues to the modulus p, and we can form a group of permutations of these numbers in the following manner.

Let s denote the operation of changing any residue z into $z+1$, and reducing the result to its least positive residue, mod p. Thus

$$s(p-1)=p, \quad s(p)=1, \quad s(1)=2, \text{ etc.,}$$

and we may write

$$s(1, 2, \ldots p)=(2, 3, \ldots p, 1).$$

Let t denote the operation of changing z into gz, and reducing the result to its least positive residue, mod p. Thus

$$t(1, 2, \ldots p)=(g, 2g, \ldots \overline{p-g}, p).$$

Evidently s is a cyclical permutation of order p; since

$$t^h(1, 2, \ldots p) \equiv (g^h, 2g^h, \ldots p),$$

and $g^h \equiv 1 \pmod p$ only when h is a multiple of $(p-1)$, it follows that t is of order $(p-1)$. It will be observed that t does not displace p, and displaces the other symbols cyclically.

It will now be proved that the $p(p-1)$ operations

$$s^m t^n \qquad \begin{bmatrix} m=1, 2, \ldots p \\ n=1, 2, \ldots (p-1) \end{bmatrix}$$

form a group.

We have
$$t^b s^c(z)=s^c(g^b z)=g^b z+c$$
$$\equiv g^b(z+l)$$

provided that
$$l \equiv c g^{p-b-1}$$

Giving l its least positive value we infer that

$$t^b s^c(z) = s^l t^b(z),$$

and
$$s^a t^b \,.\, s^c t^d = s^a \,.\, t^b s^c \,.\, t^d = s^{a+l} t^{b+d}.$$

Since a, b, c, d may be any four integers, this proves that the operations form a group. For convenience, it will be called the metacyclical group, mod p, and the reference to p may be omitted when no mistake is likely to arise.

45. There is another way of regarding the group, more convenient for some purposes, and representing the group as a set of linear substitutions. We have

$$s^x t^y(z) \equiv g^y(z+x)$$
$$\equiv lz + m,$$

provided that
$$g^y \equiv l, \quad g^y x \equiv m. \qquad (\text{mod } p)$$

If x and y are given, the last two congruences determine l, m uniquely to the modulus p. Conversely if l, m are given and l is prime to p, x and y are uniquely determined to the moduli p, $(p-1)$ respectively. Thus the group may be represented by the substitutions

$$(z,\ lz+m), \qquad \begin{bmatrix} l = 1, 2, \ldots (p-1) \\ m = 1, 2, \ldots p \end{bmatrix}$$

and in this form may be called the integral linear group.

The group is doubly transitive: that is to say, there is a definite substitution which converts any two given residues α, β into any two other given residues γ, δ. This follows from the fact that the congruences

$$l\alpha + m \equiv \gamma, \quad l\beta + m \equiv \delta \qquad (\text{mod } p)$$

admit of one and only one solution, because

$$(\alpha - \beta)\, l \equiv \gamma - \delta,$$

and $(\alpha-\beta)$, $(\gamma-\delta)$ are both prime to p.

As an example, let $p=7$, $g=3$, and let it be required to find the operation of the group which interchanges 1 and 2. The congruences

$$l + m \equiv 2, \quad 2l + m \equiv 1 \qquad (\text{mod } 7)$$

lead to $l \equiv 6$, $m \equiv 3$, and the required operation is $(z,\ 6z+3)$, or, in the other notation, $s^4 t^3$. As a verification

$$s^4(1,\ 2,\ \ldots\ 7) = (5,\ 6,\ 7,\ 1,\ 2,\ 3,\ 4),$$
$$t^3(5,\ 6,\ 7,\ 1,\ 2,\ 3,\ 4) = (2,\ 1,\ 7,\ 6,\ 5,\ 4,\ 3).$$

46. It has been shown that

$$t^b s^c = s^l t^b,$$

where l is different from c, while the index b remains *unaltered*. It follows from this that if d is any factor of $(p-1)$, including unity and $(p-1)$ itself, and if $p-1=de$, the operations

$$s^m t^{nd} \qquad \begin{bmatrix} m=1, 2, \dots p \\ n=1, 2, \dots e \end{bmatrix}$$

form a group of order pe.

This group is self-conjugate in the metacyclic group, because there is an integer i such that

$$s^a t^b \,.\, s^m t^{nd} \,.\, t^{-b} s^{-a} = s^i t^{b+nd-b} = s^i t^{nd}.$$

Let us put

$$p(p-1)=h, \quad p-1=p_1 q_1 = p_1 p_2 q_2 = \dots = p_1 p_2 \dots p_r,$$

where $p_1, p_2, \dots p_r$ are the prime factors of $(p-1)$. Then we have a composition-series

$$G_h,\ G_{pq_1},\ G_{pq_2},\ \dots\ G_{pq_{r-1}},\ G_p,\ 1,$$

with indices $$p_1, p_2, \dots p_r, p\,;$$

the notation being such that G_{pq_i} means the group of which the operations are

$$s^m t^{np_1 p_2 \dots p_i}. \qquad \begin{bmatrix} m=1, 2, \dots p \\ n=1, 2, \dots q_i \end{bmatrix}$$

In particular, G_p means the cyclical group $(1, s, s^2, \dots s^{p-1})$.

47. Suppose now that we have an equation of prime degree, and that its roots are $x_1, x_2, \dots x_p$. We obtain a group of permutations of its roots by applying to their suffixes the operations of the metacyclic group. *If this is the Galoisian group of the equation, the equation is said to be metacyclic.* An equation of this kind can be solved by a chain of auxiliaries, each cyclical and of prime degree. That the auxiliaries may be taken of prime degree follows from the composition-series just given for G_h: that they are cyclical may be inferred from the fact that they are normal as well as of prime degree, or again from the fact that $G_{pq_{i-1}} \div G_{pq_i}$ is holoedrically isomorphic with the cyclical group

$$(t^d, t^{2d}, \dots t^{(p_i-1)d}),$$

where $d=p_1 p_2 \dots p_{i-1}$ (cf. Art. 23).

48. Kronecker has put the solution of a metacyclic equation of prime degree into a very interesting form, which is analogous to that given for cyclical equations in Arts. 32–4. Before reproducing it, a few explanations and lemmas will be necessary.

As in Art. 32, we take ϵ, a primitive pth root of unity, and write

$$\theta_k = x_1 + \epsilon^k x_2 + \ldots + \epsilon^{(p-1)k} x_p. \quad [k = 1, 2, \ldots (p-1)]$$

If s, t are the generators of the metacyclic group,

$$s(\theta_k) = \Sigma \epsilon^{ki} x_{i+2} = \epsilon^{-k}\theta_k$$

as before: to find the effect of t, we observe that

$$\epsilon^k t(\theta_k) = \sum_i \epsilon^{ki} x_{gi} = \sum_i (\epsilon^{kh})^{gi} x_{gi} = \epsilon^{kh}\theta_{kh},$$

where h is determined by the congruence

$$gh \equiv 1, \qquad (\text{mod } p)$$

leading to

$$h \equiv g^{p-2} \qquad (\text{mod } p)$$

With this value of h $\quad t(\theta_k) = \epsilon^{k(h-1)}\theta_{kh}.$

It is convenient now (cf. Art. 35) to introduce a slight change of notation. We shall write

$$\vartheta_i = \theta_{g^i} \qquad [i = 0, 1, 2, \ldots (p-2)]$$

on the understanding that θ_{g^i} means θ_r, where r is the least positive residue of g^i to the modulus p. We also make the convention that for *any* positive integers m, n,

$$\vartheta_m = \vartheta_n,$$

provided that

$$m \equiv n. \qquad (\text{mod } \overline{p-1})$$

Thus there are only $(p-1)$ distinct quantities ϑ_i, and these are the same as the quantities θ_i in a different order: in particular,

$$\theta_1 = \vartheta_0 = \vartheta_{p-1}, \quad \vartheta_1 = \theta_g.$$

The effects of s and t upon ϑ_i can be found from previous formulæ: thus

$$s(\vartheta_i) = \epsilon^{-g^i}\vartheta_i,$$

$$t(\vartheta_i) = \epsilon^{g^{i-1}-g^i}\vartheta_{i-1}.$$

Let us now write

$$f_0 = \vartheta_1\vartheta_0^{-g}, \; f_1 = \vartheta_2\vartheta_1^{-g}, \; \ldots \; f_i = \vartheta_{i+1}\vartheta_i^{-g}, \; \ldots \; f_{p-2} = \vartheta_{p-1}\vartheta_{p-2}^{-g}.$$

Then $\quad s(f_i) = f_i,$

and

$$t(f_i) = \epsilon^{g^i - g^{i+1}}\vartheta_i \,.\, \epsilon^{-g(g^{i-1}-g^i)}\vartheta_{i-1}^{-g}$$

$$= \vartheta_i\vartheta_{i-1}^{-g} = f_{i-1},$$

with the special case

$$t(f_0) = f_{p-2}.$$

Consequently any rational cyclical function of $f_0, f_1, f_2, \dots f_{p-2}$ is unaltered by s and t: the quantities f_i are therefore the roots of a rational cyclic equation of degree $(p-1)$. The change of ϵ to ϵ^g converts f_i into f_{i+1}; hence it follows that when the cyclical equation aforesaid is reduced by means of the equation satisfied by ϵ, the imaginary root of unity will disappear. In other words we have *identically*, after this reduction,

$$(f-f_0)(f-f_1)\dots(f-f_{p-2}) = f^{p-1} + m_1 f^{p-2} + \dots + m_{p-1},$$

where $m_1, m_2, \dots m_{p-1}$ are *formally* metacyclic functions of $x_1, x_2, \dots x_p$, and have rational values when the given equation is metacyclic.

Suppose that we have a set of quantities $\phi_0, \phi_1, \dots \phi_{p-2}$, each of which is rational in $\epsilon, x_1, x_2, \dots x_p$ and which also satisfy the following conditions :—

(1) $s(\phi_0, \phi_1, \dots \phi_{p-2}) = \phi_0, \phi_1, \dots \phi_{p-2}$;

(2) $t(\phi_0, \phi_1, \dots \phi_{p-2}) = \phi_{p-2}, \phi_0, \phi_1, \dots \phi_{p-3}$;

(3) the change of ϵ into ϵ^g produces the same cyclical permutation as t^{-1};

(4) cyclical functions of $\phi_0, \phi_1, \dots \phi_{p-2}$ are metacyclical functions of $x_1, x_2, \dots x_p$, and can be expressed in a form which is free from ϵ.

Then by arguments precisely similar to those employed in Arts. 7, 24 it may be proved that

$$\phi_i = R(f_i), \qquad (i = 0, 1, 2, \dots \overline{p-2})$$

where R is a rational function free from ϵ, and the coefficients of the powers of f_i are metacyclic functions of $x_1, x_2, \dots x_p$.

49. From the equations which express the quantities f_i in terms of the quantities ϑ_i we can eliminate all the ϑ's except ϑ_0 in the following manner. Raise the first equation to the power g^{p-2}, the second to the power g^{p-3}, etc., and multiply all the results together: observing that $\vartheta_{p-1} = \vartheta_0$, we have

$$\vartheta_0^{1-g^{p-1}} = f_0^{g^{p-2}} f_1^{g^{p-3}} \dots f_i^{g^{p-i-2}} \dots f_{p-3}^g f_{p-2} \quad \dots\dots\dots(1).$$

The primitive root g may always be chosen in such a way that

$$g^{p-1} - 1 = p(kp-1),$$

where k is a positive integer. Supposing this done,

$$\vartheta_0^{1-g^{p-1}} = \vartheta_0^p \div (\vartheta_0^{kp})^p.$$

Now the quantities $\vartheta_0^{kp}, \vartheta_1^{kp}, \ldots \vartheta_{p-2}^{kp}$ satisfy all the conditions enumerated in the latter part of Art. 48, so that we may put

$$\vartheta_0^{kp} = R(f_0) \quad \text{.........................(2)},$$

where R is a rational function of the nature explained above.

The positive integers $r_1, r_2, \ldots r_{p-2}$ can be uniquely determined so that

$$g^{p-2} = q_{p-2}p + r_{p-2}, \quad g^{p-3} = q_{p-3}p + r_{p-3}, \text{ etc.}$$

with $$0 < r_i < p \qquad (i = 1, 2, \ldots \overline{p-2})$$

and the quantities q_i positive integers or zeros.

If, now, we write, as an abbreviation,

$$K_0 = R(f_0)\, f_0^{q_{p-2}} f_1^{q_{p-3}} \ldots f_{p-3}^{q_1} \quad \text{..................(3)},$$

we obtain from (1), after multiplying both sides by $\vartheta_0^{kp^2}$,

$$\vartheta_0^p = K_0^p f_0^{r_{p-2}} f_1^{r_{p-3}} \ldots f_{p-3}^{r_1} f_{p-2} \quad \text{..............(4)}.$$

From this it follows that

$$\vartheta_i^p = K_i^p f_i^{r_{p-2}} f_{i+1}^{r_{p-3}} \ldots f_{i+p-2} \quad \text{..................(5)},$$

where K_i is derived from K_0 by changing $f_0, f_1, f_2, \ldots$ into $f_i, f_{i+1}, f_{i+2}, \ldots$ respectively.

The relations (1), (2), (4), (5) are all reducible to identities, whatever $x_1, x_2, \ldots x_p$ may be, solely in virtue of the equation satisfied by ϵ, and the definitions of ϑ_i, f_i, etc. If $x_1, x_2, \ldots x_p$ are the roots of a metacyclic equation with numerical coefficients, $f_0, f_1, \ldots f_{p-2}$ are the roots of an auxiliary cyclical equation with rational coefficients. By the adjunction of f_0 the other roots become rational, and finally, if we put

$$\tau_i = \sqrt[p]{f_i}$$

a definite pth root of f_i, we have

$$px_1 = -c_1 + \Sigma\vartheta_i = -c_1 + \sum_i K_i \tau_i^{r_{p-2}} \tau_{i+1}^{r_{p-3}} \ldots \tau_{i+p-3}^{r_1} \tau_{i+p-2}.$$

If, in the expression on the right, we give to each quantity τ_i any one of its p different values, we only obtain p different expressions on the whole: thus the formula may be used to determine any root of the given equation, and it does not lead to any value of x_1 which is not a root.

50. When $p = 3$, the metacyclic group consists of all the permutations of three things: hence the general cubic equation is metacyclic. To solve it by Kronecker's method we take $g = 5$,

$$\vartheta_0 = \alpha + \omega\beta + \omega^2\gamma, \qquad \vartheta_1 = \alpha + \omega^2\beta + \omega\gamma,$$
$$f_0 = \vartheta_1\vartheta_0^{-5}, \qquad f_1 = \vartheta_0\vartheta_1^{-5}.$$

With the notation of Art. 4, we find that f_0, f_1 are the roots of

$$B^5 f^2 - (A^2 - 2B^3) f + B = 0 \quad \text{.................(1).}$$

Moreover
$$f_0^5 f_1 = \vartheta_0^{-24},$$
$$\vartheta_0^3 = f_0^5 f_1 (\vartheta_0^3)^9 \quad \text{.....................(2),}$$

and we have now to express ϑ_0^3 in terms of f_0. To do this by the general method is a good exercise; but it is simpler to proceed as follows. We have

$$f_1 - f_0 = \frac{(\vartheta_0^3 + \vartheta_1^3)(\vartheta_0^3 - \vartheta_1^3)}{(\vartheta_0\vartheta_1)^5} = \frac{A}{B^5}(\vartheta_0^3 - \vartheta_1^3);$$

hence
$$A(\vartheta_0^3 - \vartheta_1^3) = B^5(f_1 - f_0) = B^5(f_0 + f_1 - 2f_0)$$
$$= A^2 - 2B^3 - 2B^5 f_0,$$

and
$$\vartheta_0^3 = \tfrac{1}{2}(\vartheta_0^3 - \vartheta_1^3 + A) = \frac{A^2 - B^3 - B^5 f_0}{A} \quad \text{............(3).}$$

If we write
$$f_0 = \tau_0^3, \qquad f_1 = \tau_1^3$$

we obtain from (2) and (3)

$$\vartheta_0 = \frac{f_0(A^2 - B^3 - B^5 f_0)^3}{A^3}\tau_0^2\tau_1, \qquad \vartheta_1 = \frac{f_1(A^2 - B^3 - B^5 f_1)^3}{A^3}\tau_1^2\tau_0.$$

To put the solution into its simplest form, we must express the multipliers of $\tau_0^2\tau_1$ and $\tau_1^2\tau_0$ as linear functions of f_0 and f_1 respectively. The final result is

$$\vartheta_0 = \frac{B(B^5 f_0 - A^2 + B^3)}{A}\tau_0^2\tau_1,$$

$$\vartheta_1 = \frac{B(B^5 f_1 - A^2 + B^3)}{A}\tau_1^2\tau_0,$$

$$3x = -c_1 + \vartheta_0 + \vartheta_1.$$

This gives the solution in a definite form whenever the values of A and B are both different from zero. When $A = 0$, the expressions for ϑ_0 and ϑ_1 assume the indeterminate form 0/0: in this special case the cubic has the rational root $-c_1/3$, and the others are the roots of a rational quadratic. When $B = 0$ the cubic may be written

$$(3x + c_1)^3 + c_1^3 - 27c_3,$$

and is cyclical. Finally, when $A = B = 0$ the cubic has three equal roots.

QUINTIC EQUATIONS

51. It is an interesting problem to find the most general form of a metacyclic equation of the fifth degree. To do this, we must first find the most general form of a cyclic quartic.

The roots of such an equation are, with $m = a^2 + b^2$, so that

$$\sqrt{(m + a\sqrt{m})} = \frac{a + \sqrt{m}}{b}\sqrt{(m - a\sqrt{m})},$$

$$\begin{aligned} x_1 &= c + d\sqrt{m} + (e + g\sqrt{m})\sqrt{(m + a\sqrt{m})} = h(x_4), \\ x_2 &= c - d\sqrt{m} + (e - g\sqrt{m})\sqrt{(m - a\sqrt{m})} = h(x_1), \\ x_3 &= c + d\sqrt{m} - (e + g\sqrt{m})\sqrt{(m + a\sqrt{m})} = h(x_2), \\ x_4 &= c - d\sqrt{m} - (e - g\sqrt{m})\sqrt{(m - a\sqrt{m})} = h(x_3). \end{aligned}$$

Elimination of the radicals leads to the most general cyclic quartic in the form

$$\begin{aligned} (x - c)^4 &- 2m(d^2 + e^2 + mg^2 + 2aeg)(x - c)^2 \\ &- 4md\{a(e^2 + mg^2) + 2meg\}(x - c) + m^2d^4 \\ &- 2m^2d^2(e^2 + mg^2 + 2aeg) + mb^2(e^2 - mg^2)^2 = 0. \end{aligned}$$

Now let k, l, n, p, q be any rational quantities; and let

$$x_i = \tau_i^5, \qquad (i = 1, 2, 3, 4)$$

$$f(x) = kx^3 + lx^2 + nx + p,$$

$$\begin{aligned} \xi = f(x_1)\,\tau_1^3\tau_2^4\tau_3^2\tau_4 + f(x_2)\,\tau_2^3\tau_3^4\tau_4^2\tau_1 \\ + f(x_3)\,\tau_3^3\tau_4^4\tau_1^2\tau_2 + f(x_4)\,\tau_4^3\tau_1^4\tau_2^2\tau_3 + q. \end{aligned}$$

Then x_1, x_2, x_3, x_4 being the roots of a cyclic quartic as previously constructed, ξ will be a root of a rational quintic which is metacyclic in the field $(a, b, c, d, e, k, l, n, p, q)$.

It is supposed here that the notation for the roots of the quartic is so arranged that its Galoisian group consists of the cyclical permutation $(x_1x_2x_3x_4)$ and its powers. This having been done, we may give *each* of the quantities τ_i all its five values, without obtaining more than five values for ξ. There will generally be five different values: but there may be repetitions for particular values of $(a, b, \ldots q)$.

52. The general quintic can be transformed, with the help of solvable equations, to the standard form

$$x^5 + \alpha x + \beta = 0 \quad\text{..........................(1)},$$

and if this is metacyclic its roots can be actually found in the following manner.

The generators of the metacyclic group may be taken to be

$$s = (12345), \quad t = (1243)(5);$$

and if we put $\epsilon = e^{2\pi i/5}$,

$$\phi = x_1 x_2 + x_2 x_3 + x_3 x_4 + x_4 x_5 + x_5 x_1,$$

it is found by actual calculation that, in virtue of $\Sigma x_i = \Sigma x_i^2 = \Sigma x_i^3 = 0$,

$$\theta_1 \theta_4 = \phi \sqrt{5} = -\theta_2 \theta_3 \quad \text{.........................(2)},$$

$$\theta_1^2 \theta_3 + \theta_2^2 \theta_1 + \theta_3^2 \theta_4 + \theta_4^2 \theta_2 = 0 \quad \text{.................(3)},$$

θ_1, θ_2, etc. having the same meaning as in Art. 48 and elsewhere. If we write, for simplicity,

$$\phi \sqrt{5} = u$$

and eliminate θ_3, θ_4 from (2) and (3), the result may be written in the form

$$u^3 \theta_1^{10} + u^2 (\theta_1^3 \theta_2)^3 + (\theta_1^3 \theta_2)^4 - u \theta_1^{10} (\theta_1^3 \theta_2) = 0.$$

This is satisfied identically, and in the most general manner, by putting

$$\left.\begin{aligned} u &= l(l^2 - 1) t^2 \\ \theta_1^5 &= l^4 (l+1)^3 (l-1)^2 t^5 \\ \theta_1^3 \theta_2 &= l^3 (l^2 - 1)^2 t^4 \end{aligned}\right\} \quad \text{....................(4)},$$

l and t representing two independent parameters.

Now one root of the quintic is given by

$$\begin{aligned} 5x &= \theta_1 + \theta_2 + \theta_3 + \theta_4 \\ &= \theta_1 + \frac{l^3 (l^2 - 1)^2 t^4}{\theta_1^3} - \frac{\theta_1^3}{l^2 (l^2 - 1) t^2} + \frac{l(l^2 - 1) t^2}{\theta_1} \quad \text{.........(5)}, \end{aligned}$$

by means of (2) and (4). Eliminating θ_1 from this and the second of equations (4), we find that

$$\begin{aligned} 5^5 x^5 - l(l^2 - 1) t^4 [25 (l^2 + l - 1)(l^2 - 4l - 1) x \\ + (l^2 + 1)(l^4 + 22l^3 - 6l^2 - 22l + 1) t] = 0 \quad \text{......(6)}. \end{aligned}$$

It will now be supposed that l and t have values such that the equations (6) and (1) are equivalent: thus

$$l(l^2 - 1)(l^2 + l - 1)(l^2 - 4l - 1) t^4 + 125\alpha = 0 \quad \text{......(7)},$$

$$l(l^4 - 1)(l^4 + 22l^3 - 6l^2 - 22l + 1) t^5 + 3125\beta = 0 \quad \text{......(8)}$$

It remains to make use of the fact that (1) is metacyclic. The substitution s makes no change in ϕ, and in virtue of $\Sigma x_i x_j = 0$ the substitution t converts ϕ into $-\phi$: consequently ϕ^2 is a metacyclic function, and its value is rational. Denoting it by γ, we deduce from the first of equations (4)

$$l^2 (l^2 - 1)^2 t^4 = 5\gamma \quad \text{........................(9)},$$

and from this and (7)

$$\gamma(l^2+l-1)(l^2-4l-1)+25a(l^2-1)l=0.$$

The solution of this is given by

$$\left.\begin{aligned}\gamma y^2+(25a-3\gamma)y-4\gamma=0\\ l^2-yl-1=0\end{aligned}\right\} \quad\text{..................(10).}$$

From (7) and (8)

$$t=\frac{25(l^2+l-1)(l^2-4l-1)\beta}{(l^2+1)(l^4+22l^3-6l^2-22l+1)a} \quad\text{............(11),}$$

and from (4)

$$\theta_1^5=\frac{5^{10}(l^2+l-1)^5(l^2-4l-1)^5(l+1)^3(l-1)^2l^4\beta^5}{(l^2+1)^5(l^4+22l^3-6l^2-22l+1)^5a^5} \quad\text{......(12).}$$

Equations (10), (11), (12) and (5) contain the complete solution of the problem, supposing that the value of γ is known; and it will be observed that, in accordance with theory, the degrees of the auxiliary equations are 2, 2 and 5, the prime factors of the order of the metacyclic group.

The quantity γ is a root of the equation*

$$(\gamma-a)^4(\gamma^2-6a\gamma+25a^2)=5^5\beta^4\gamma \quad\text{..............(13),}$$

so that the quintic is, or is not, metacyclic in any given field according as (13) has or has not a rational root in that field. If the field is (a, β), we must have rational quantities λ, μ such that

$$\gamma=\lambda a, \quad \beta=\mu a;$$

whence

$$a=\frac{5^5\mu^4\lambda}{(\lambda-1)^4(\lambda^2-6\lambda+25)}, \quad \beta=\frac{5^5\mu^5\lambda}{(\lambda-1)^4(\lambda^2-6\lambda+25)}.$$

It may be observed that the solution of (6) assumes a very elegant form if we put

$$l=\wp(z)$$

where $\wp(z)$ is a lemniscate function of z; that is to say, one for which

$$g_3=0.$$

53. The condition that a general quintic may be metacyclic has been put into an invariant form by Mr W. E. H. Berwick, and, with his permission, the result is given here. Let J, K, L be the invariants which Salmon denotes by those letters (*H. A.*, pp. 228–31); and let

$$j=5^8.J, \quad k=2^5.5^6.K, \quad l=-2^{10}.5^9.L.$$

* C. Runge, *Acta Math.* 7.

If $x_1, x_2, \dots x_5$ are the roots of the quintic $(a, b, c, d, e, f \between x, 1)^5 = 0$, and we put

$$\phi = a^4 \{(x_1 - x_2)^2 (x_2 - x_3)^2 (x_3 - x_4)^2 (x_4 - x_5)^2 (x_5 - x_1)^2 + (x_1 - x_3)^2 (x_3 - x_5)^2 (x_5 - x_2)^2 (x_2 - x_4)^2 (x_4 - x_1)^2\},$$

then ϕ is a metacyclic function of the roots and satisfies the equation

$$\begin{aligned}\phi^6 &+ 10j\phi^5 + (35j^2 + 10k)\phi^4 \\ &+ (60j^3 + 30jk + 10l)\phi^3 \\ &+ (55j^4 + 30j^2k + 25k^3 + 50jl)\phi^2 \\ &+ (26j^5 + 10j^3k + 44jk^2 + 59j^2l + 14kl)\phi \\ &+ (5j^6 + 20j^2k^2 + 20j^3l + 20jkl + 25l^2) = 0.\end{aligned}$$

So the required condition is that this sextic must have a rational root.

This resolvent can also be readily used to distinguish the sub-groups

$$\Gamma_{10} = \{s, t^2\}, \quad \Gamma_5 = \{s\}$$

of the metacyclic group. Putting

$$\chi_1 = a^4 (x_1 - x_2)^2 (x_2 - x_3)^2 (x_3 - x_4)^2 (x_4 - x_5)^2 (x_5 - x_1)^2,$$
$$\chi_2 = a^4 (x_1 - x_3)^2 (x_3 - x_5)^2 (x_5 - x_2)^2 (x_2 - x_4)^2 (x_4 - x_1)^2,$$

when $\phi \equiv \chi_1 + \chi_2$ is a rational root of the resolvent, χ_1 and χ_2 are the roots of

$$\chi^2 - \phi\chi + \Delta = 0.$$

This quadratic must have rational roots when the group reduces to Γ_{10}, and it further reduces to Γ_5 when χ_1 or χ_2 is a rational square.

SEXTIC EQUATIONS

54. There are five main classes of irreducible sextic equation, falling into sixteen sub-classes due to the transitive groups:

I. H_{720}, Γ_{360}.
II. H_{120}, Γ_{60}.
III. G_{72}, Γ_{36}, G_{36}, G_{18}.
IV. G_{48}, G_{24}, H_{24}, Γ_{24}, Γ_{12}.
V. G_{12}, H_6, C_6.

In this scheme the first group in each line contains the others as sub-groups, Γ_r contains even permutations only, C_6 is cyclical and H_r is simply isomorphic with the symmetric group of degree r. G_{12} is a sub-group of both G_{72} and G_{48}, while Γ_{360}, H_{120}, G_{72}, G_{48} are the only transitive maximum sub-groups of H_{720}. Due to these maximum sub-groups there are four *principal resolvents* of degrees 2, 6, 10, 15. At least one principal resolvent has a rational root when the group of the sextic is unsymmetric.

I. H_{720} and Γ_{360}. *The General Sextic.*

When no one of the resolvents of degrees 6, 10, 15 has a rational root, the sextic is general in character, its group being H_{720} or Γ_{360} according as Δ is not or is a rational square.

55. Before examining the other classes of sextic equation in detail it is necessary to prove two theorems on the structure of the roots of an irreducible equation $a(x)=0$ of any degree n. A principal resolvent $\Theta(\theta)=0$ of $a(x)=0$, due to the maximum sub-group M, is, in general, reducible when the group G of the equation is a sub-group of M. An irreducible factor $\Theta_1(\theta)$ of $\Theta(\theta)$ will be called an *irreducible resolvent* for G.

We take $\alpha, \beta, \ldots \kappa$ to be the roots of $a(x)=0$ and $\theta_1, \theta_2, \ldots \theta_p$ rational functions of these roots to be the zeros of an irreducible resolvent for the group G of order nh. Each permutation P of G corresponds to a unique permutation Q in the r θ's and the latter permutations may or may not be all distinct.

THEOREM (1). When no two of the nh permutations Q are identical, these permutations form a group G' simply isomorphic with G. The permutations of G which do not change α form a sub-group G_α, and the h equivalent permutations in the θ's define a group G_α' of the same order h. A symmetric function $\Theta(\theta_1, \theta_2, \ldots \theta_r)$ of the linear forms into which

$$c_1\theta_1 + c_2\theta_2 + \ldots + c_r\theta_r$$

is changed by the permutations of G_α', when expressed in terms of $\alpha, \beta, \ldots \kappa$, is unaltered by the permutations of G_α. The function Θ can always be so chosen as to be different in value from its conjugates. In these circumstances α and Θ belong to the same group G_α and

$$\Theta = R_1(\alpha),\ \alpha = R_2(\Theta) = R(\theta_1, \theta_2, \ldots \theta_r),$$

a rational function unaltered in form by the permutations of G_α'. Applying the permutation

$$\begin{Bmatrix} \alpha, \ldots, \theta_1, \ldots \theta_r \\ \beta, \ldots, \theta_a, \ldots \theta_c \end{Bmatrix}$$

(a member of G) to the last equation, it follows that

$$\beta = R(\theta_a, \ldots \theta_c).$$

Hence, when the groups G, G' are simply isomorphic, the roots $\alpha, \beta, \ldots \kappa$ are expressible rationally in terms of those of the irreducible resolvent.

THEOREM (2). When G is multiply isomorphic with G', the permutations of G which leave $\theta_1, \theta_2, \ldots \theta_r$ unaltered in order are $1, S_2, \ldots S_{h_1}$, forming a sub-group G_1 of G. If T' is a permutation of G not in G_1,

$$T'S_i(\theta_1, \ldots \theta_r) = T'(\theta_1, \ldots \theta_r) = (\theta_a, \ldots \theta_e),$$
$$S_iT'(\theta_1, \ldots \theta_r) = S_i(\theta_a, \ldots \theta_e) = (\theta_a, \ldots \theta_e).$$

Hence the set of permutations $T'G_1$ is identical with the set G_1T', though not necessarily in the same order: in fact, if $h_1k_1 = nh$, all the permutations of G fall into the scheme

$$G_1,\ T_2'G_1, \ldots T_{\kappa_1}'G_1.$$

Another irreducible resolvent $\Lambda(\lambda) = 0$, for which G_2, T_j, h_2, k_2 are defined in the same way as G_1, S_i, h_1, k_1 for $\Theta(\theta)$, can often be chosen in such a way that G_1, G_2 have no common permutation except identity. In these circumstances no two of the permutations of G_1 derange the sequence $(\lambda_1, \lambda_2, \ldots \lambda_s)$ in the same way. For if

$$S_i(\lambda_1, \lambda_2, \ldots \lambda_s) = S_\kappa(\lambda_1, \lambda_2, \ldots \lambda_s) = (\lambda_u, \lambda_v, \ldots \lambda_t),$$

then $(S_\kappa^{-1}S_i)(\lambda_1, \lambda_2, \ldots \lambda_s) = S_\kappa^{-1}(\lambda_u, \lambda_v, \ldots \lambda_t) = (\lambda_1, \lambda_2, \ldots \lambda_s)$, showing $S_\kappa^{-1}S_i$ to be a member of G_2, *i.e.* $S_i = S_\kappa$. Thus every member of G_1 permutes the λ's differently, whence

$$h_1 \leqslant k_2 = nh/h_2 \quad \text{or} \quad h_1h_2 \leqslant nh.$$

In the case in which $h_1h_2 = nh$, $h_2 = k_1$ and every member of the group $\{T_j\}$ permutes the θ's differently. Hence $T_j = T_j'S_i$, and the set T_jG_1 is identical with $T_j'G_1$ or G_1T_j' or G_1T_j, *i.e.*

$$G = G_1 + T_2G_1 + \ldots + T_{h_2}G_1$$
$$= G_2 + S_2G_2 + \ldots + S_{h_1}G_2.$$

The permutation T_jS_i is in the set T_jG_1, also in the set S_iG_2. Similarly S_iT_j appears in each of these sets. The several members of T_jG_1 appear in the h_1 different sets S_iG_2, and so for the members of S_iG_2. Accordingly T_jS_i, S_iT_j can only appear in both when these permutations are identical. Hence, whenever G_1, G_2 have no common permutation except identity, and the product of their orders is the order of G, the group G is the direct product of G_1 and G_2.

It is also readily proved that $\alpha, \beta, \ldots \kappa$ are expressible rationally in terms of θ's and λ's. The permutations of G which leave α unchanged form a group G_α of order h. Applying these permutations to a general rational function of θ's and λ's, there are h such functions different in form and numerical value. A symmetric function of these belongs to the group G_α, hence

$$\alpha = S_\alpha(\theta_1, \ldots \theta_r, \lambda_1, \ldots \lambda_s),$$

a rational function unaltered in form by the permutations of G_α.

In the application to the sextic it is unnecessary to enter into further results (i) when G_1, G_2 have no common permutation except identity and $h_1 h_2 < nh$, or (ii) when G_1, G_2 have a common sub-group.

56. II. H_{120} and Γ_{60}. *Sextic Resolvents of a Quintic.*

If $$S=(\alpha\beta\gamma\delta),\quad T=(\alpha\zeta\gamma\delta\beta),\quad U=(\alpha\beta\epsilon\gamma\zeta\delta),$$
the permutations of H_{120} are
$$S^l T^m U^n \text{ or } U^n S^l T^m.$$
These permutations leave unaltered a symmetric function of
$$\alpha\beta+\gamma\epsilon+\delta\zeta,\quad \alpha\gamma+\delta\beta+\zeta\epsilon,\quad \alpha\delta+\zeta\gamma+\epsilon\beta,\quad \alpha\zeta+\epsilon\delta+\beta\gamma,\quad \alpha\epsilon+\beta\zeta+\gamma\delta.$$
To form a resolvent it is better to adopt the invariant function
$$\begin{aligned}\phi_6 = a^2\{&(\alpha-\beta)^2(\gamma-\epsilon)^2(\delta-\zeta)^2\\ &+(\alpha-\gamma)^2(\delta-\beta)^2(\zeta-\epsilon)^2\\ &+(\alpha-\delta)^2(\zeta-\gamma)^2(\epsilon-\beta)^2\\ &+(\alpha-\zeta)^2(\epsilon-\delta)^2(\beta-\gamma)^2\\ &+(\alpha-\epsilon)^2(\beta-\zeta)^2(\gamma-\delta)^2\}\end{aligned}$$
whose conjugates are
$$\phi_1=\phi_6(\gamma\alpha),\quad \phi_2=\phi_6(\alpha\epsilon),\quad \phi_3=\phi_6(\epsilon\gamma),\quad \phi_4=\phi_6(\alpha\epsilon\gamma),\quad \phi_5=\phi_6(\alpha\gamma\epsilon).$$
The sextic resolvent satisfied by the six ϕ's, as calculated by Mr C. W. Gilham, is
$$\begin{aligned}\Phi(\phi) &= \phi^6 + 6c_2\phi^5 + (15c_2^2 + 8c_4)\phi^4\\ &+ (20c_2^3 + 30c_2c_4 - 2c_6)\phi^3 + (15c_2^4 + 42c_2^2c_4 - 6c_2c_6 + 16c_4^2)\phi^2\\ &+ (6c_2^5 + 26c_2^3c_4 - 6c_2^2c_6 + 24c_2c_4^2 - 8c_4c_6 - 4c_{10})\phi\\ &+ (c_2^6 + 6c_2^4c_4 - 2c_2^3c_6 + 9c_2^2c_4^2 - 6c_2c_4c_6 - 3c_2c_{10} + c_6^2) = 0,\end{aligned}$$
where c_2, c_4, c_6, c_{10} are invariants of the binary sextic form $a\Pi(x-\alpha y)$.

For H_{120} and Γ_{60} $\Phi(\phi)$ has a rational linear factor $\phi-\phi_6$, the remaining quintic factor being irreducible. H_{120} is simply isomorphic with the symmetric group $\{\phi_1\phi_2\phi_3\phi_4\phi_5\}$ and Theorem 1 of § 55 applies. The twenty permutations which leave α unaltered are V_1^m, W_1^m, where
$$V_1=(\gamma\epsilon\delta\beta\zeta)=(\phi_1\phi_2\phi_5\phi_3\phi_4),$$
$$W_1=(\gamma\epsilon\beta\delta)=(\phi_1\phi_2\phi_3\phi_5).$$
Hence $\alpha = M(\phi_1, \phi_2, \phi_5, \phi_3, \phi_4)$, a rational metacyclic function, and the remaining roots of the sextic are similarly expressed on applying the permutation
$$(\alpha\beta\epsilon\gamma\zeta\delta)=(\phi_1\phi_2\phi_4)(\phi_3\phi_5).$$

57. III. G_{72}, Γ_{36}, G_{36} and G_{18}. *Bicubic Sextics.*

The ten-valued function $a\,(\alpha\epsilon\gamma+\delta\beta\zeta)$ belongs to the group

$$G_{72}=(\{\alpha\epsilon\gamma\},\{\delta\beta\zeta\})\,\{(\alpha\delta)(\epsilon\beta)(\gamma\zeta)\}.$$

Mr Gilham has calculated the resolvent $\mathrm{X}(\chi)=0$ satisfied by the more convenient function

$$\begin{aligned}\chi=\tfrac{1}{4}a^2\{&(\alpha-\delta)(\epsilon-\beta)(\gamma-\zeta)+(\alpha-\delta)(\epsilon-\zeta)(\gamma-\beta)\\&+(\alpha-\beta)(\epsilon-\zeta)(\gamma-\delta)+(\alpha-\zeta)(\epsilon-\beta)(\gamma-\delta)\\&+(\alpha-\zeta)(\epsilon-\delta)(\gamma-\beta)+(\alpha-\beta)(\epsilon-\delta)(\gamma-\zeta)\}^2.\end{aligned}$$

When $\mathrm{X}(\chi)=0$ has a rational root $\chi_{10}=mk^2$ the functions

$$(\alpha+\epsilon+\gamma-\delta-\beta-\zeta)^2,\quad(\alpha+\epsilon+\gamma-\delta-\beta-\zeta)/\chi_{10},$$
$$(\epsilon\gamma+\gamma\alpha+\alpha\epsilon-\beta\zeta-\zeta\delta-\delta\beta)/\chi_{10},\quad(\alpha\epsilon\gamma-\delta\beta\zeta)/\chi_{10}$$

which belong to G_{72} take rational values, and the sextic is

$$(ax^3+3bx^2+3c'x+d')^2-m\,(c'x^2+2f'x=g')^2,$$

thus breaking into cubic factors in $[\sqrt{m}]$.

The permutations of G_{36} derange ϕ_1, ϕ_2, ϕ_3 *inter se*, also ϕ_4, ϕ_5, ϕ_6 *inter se*. So for this group $\Phi(\phi)$ is the product of irreducible cubic factors. All the derangements of the six ϕ's are different: in fact G_{36} is simply isomorphic with $\{\phi_1\phi_2\phi_3\}\,\{\phi_4\phi_5\phi_6\}$ and Theorem 2 of § 55 applies. The functions

$$\alpha,\ \phi_1\phi_4+\phi_2\phi_5+\phi_3\phi_6,$$

being unaltered by the same six permutations of G_{36},

$$\alpha=R\,(\phi_1\phi_4+\phi_2\phi_5+\phi_3\phi_6),$$

and, applying the permutations of $\{\phi_1\phi_2\phi_3\}$, ϵ, γ, etc. are the same rational functions of

$$\phi_i\phi_4+\phi_j\phi_5+\phi_k\phi_6,\quad i,j,k=1,2,3.$$

The roots of a G_{36}-sextic are thus expressible rationally in terms of the roots of two irreducible cubics. Such an equation remains irreducible, however, when the field of its coefficients is enlarged by adjunction of one (or all) the roots of either of the cubics. An equation of this type is termed *compound*.

58. IV. G_{48}, G_{24}, H_{24}, Γ_{24}, Γ_{12}. *Cubiquadratic Sextics.*

The fifteen-valued function

$$\psi=\alpha\delta+\epsilon\beta+\gamma\zeta$$

belongs to the last maximum sub-group

$$G_{48}=\{(\alpha\delta)\}\,\{(\alpha\epsilon\delta\beta)\}\,\{(\alpha\beta\gamma\delta\epsilon\zeta)\}.$$

When the 15-ic in ψ has a rational root the sextic breaks up into the quadratic factors

$$a\{x^2+f_1(\theta)x+f_2(\theta)\}\{x^2+f_1(\theta')x+f_2(\theta')\}\{x^2+f_1(\theta'')x+f_2(\theta'')\},$$

$\theta, \theta', \theta''$ being conjugate cubic irrationalities. Hence

$$\alpha=R(\theta, \sqrt{h}(\theta)), \quad \delta=R(\theta, -\sqrt{h}(\theta)), \text{ etc.}$$

For each of the five groups ϕ_5 and ϕ_6 are rational, the remaining quartic factor of $\Phi(\phi)$ being irreducible. It is known that the roots of such a quartic are given by

$$\phi_1-\phi_2-\phi_3+\phi_4=\sqrt{k}(\theta_1), \text{ etc.,}$$

where $\theta_1, \theta_1', \theta_1''$ are the roots of the reducing cubic. The functions

$$\alpha \text{ and } (\phi_1-\phi_2-\phi_3+\phi_4)\sqrt{\Delta}$$

are undisturbed by the same eight permutations of G_{48}, hence

$$\alpha=R_0[(\phi_1-\phi_2-\phi_3+\phi_4)\sqrt{\Delta}],$$
$$\delta=R_0[(-\phi_1+\phi_2+\phi_3-\phi_4)\sqrt{\Delta}], \text{ etc.,}$$

shewing that the G_{48}-sextic defines another type of compound irrationality.

59. V. G_{12}, H_6, C_6. *Composite Sextic Equations.*

For these groups the 10-ic and 15-ic resolvents have each a rational root. Hence the quadratic factors of IV and the cubic factors of III both exist, and it follows from the Euclidean G.C.M. process that

$$\alpha=R_1(\theta, \sqrt{m}), \quad \delta=R_1(\theta, -\sqrt{m}).$$

The group reduces to H_6 when

$$m(\theta-\theta')^2(\theta-\theta'')^2(\theta'-\theta'')^2$$

is a rational square, whence $\alpha=R_2(\theta, \theta', \theta'')$.

Finally it reduces to the cyclical group C_6 when $[\theta]$ is normal.

CHAPTER V

SOLUTION BY STANDARD FORMS

60. As explained in Chap. I (Art. 27), the first step towards the solution of an equation, after determining its Galoisian group, is to construct a series of Galoisian auxiliaries. If the degree of each auxiliary is prime, the equation is solvable by radicals, because each auxiliary is cyclical; and it can be proved that in no other case is the original equation solvable by radicals. The group of each auxiliary is simple; hence the only outstanding difficulty is the discussion of non-cyclical equations, of which the Galoisian groups are simple. The reason why the general equation of order n cannot be solved algebraically when $n > 4$ is that the group of even permutations of n things is simple* except when $n = 4$. The cases $n = 2$ and $n = 3$ are also exceptional, because in the first case there are no even permutations, and in the second they form a cyclical group of order 3.

The most effective way of attacking an equation of which the group is non-cyclical and simple is to transform it, if possible, into another equation of standard form, for which the solution is known or has been tabulated. The spirit of the method may be illustrated, in the first place, by considering the cubic equation

$$x^3 \pm ax \pm b = 0,$$

where a, b denote real positive quantities. If we put

$$x = ky, \quad 3k^2 = 4a, \quad c = 4b/k^3$$

the equation becomes

$$4y^3 \pm 3y \pm c = 0;$$

and by properly choosing the sign of k, we can make this

$$4y^3 \pm 3y - c = 0,$$

with $c > 0$. If the coefficient of y is -3, and c is a proper fraction, we may find a real quantity θ such that $\cos 3\theta = c$, and then

$$y = \cos\theta, \ \cos\left(\theta + \frac{2\pi}{3}\right), \ \cos\left(\theta + \frac{4\pi}{3}\right);$$

* Burnside, *Theory of Groups*, p. 153.

while if $c > 1$ we find θ such that $\cosh 3\theta = c$, and then

$$y = \cosh\theta,\ \cosh\left(\theta + \frac{2\pi i}{3}\right),\ \cosh\left(\theta + \frac{4\pi i}{3}\right).$$

On the other hand, if the coefficient of y is $+3$, we may find θ such that $\sinh 3\theta = c$, and then

$$y = \sinh\theta,\ \sinh\left(\theta + \frac{2\pi i}{3}\right),\ \sinh\left(\theta + \frac{4\pi i}{3}\right).$$

Thus in every case the equation is solved with the help of a table of trigonometrical or hyperbolic functions.

61. Several methods of this kind, all indeed ultimately equivalent, have been applied to the general quintic. One of these, the solution by means of the icosahedral irrationality, will now be given in outline; for further details the reader is referred to Klein's lectures on the icosahedron, and to the treatise on modular functions by Klein and Fricke.

A point on a sphere may be determined by its north polar distance θ and longitude ϕ. If we put

$$z_1 : z_2 = \tan\frac{\theta}{2}(\cos\phi + i\sin\phi),$$

z_1, z_2 may be taken as homogeneous coordinates defining the position of the point. Suppose, now, that we have a regular icosahedron inscribed in the sphere, with one vertex at the point $\theta = 0$ and another on the great circle $\phi = 0$. If we put

$$f = z_1 z_2 (z_1^{10} + 11 z_1^5 z_2^5 - z_2^{10})$$

the roots of $f = 0$ correspond to the twelve vertices of the solid. The binary form f has two covariants

$$H = -(z_1^{20} + z_2^{20}) + 228(z_1^{15} z_2^5 - z_1^5 z_2^{15}) - 494 z_1^{10} z_2^{10},$$

$$T = (z_1^{30} + z_2^{30}) + 522(z_1^{25} z_2^5 - z_1^5 z_2^{25}) - 10005(z_1^{20} z_2^{10} + z_1^{10} z_2^{20}),$$

and the three forms are connected by the identity

$$H^3 + T^2 = 1728 f^5.$$

The roots of $H = 0$ correspond to the centres of the equilateral triangles into which the surface of the sphere is divided by the great circle arcs into which the edges of the icosahedron are projected from the centre of the sphere; and the roots of $T = 0$ correspond to the middle points of the sides of these triangles. H is the Hessian of f, and T is the Jacobian of H and f.

Let AB be a side of any one of the 20 triangles, and CD any other of the remaining 29 sides. Then there is a definite rotation about a diameter of the sphere which brings AB into coincidence with CD. Similarly there is a definite rotation which brings AB into coincidence with DC. We thus obtain 58 rotations, each of which, applied to the icosahedron, brings it into a new position in which it occupies the same space as before. Besides these, there is the rotation about the diameter bisecting AB, which brings AB into coincidence with BA. Altogether, there are sixty different positions of the icosahedron, and if we include, as the identical operation, that of leaving the icosahedron alone, we have a group of 60 rotations which form a group. Each rotation may be associated with a linear substitution applied to z_1 and z_2. If we put

$$e^{2\pi i/5} = \epsilon,$$

$$s(z_1, z_2) = (\epsilon^3 z_1, \epsilon^2 z_2), \quad t(z_1, z_2) = \left(\frac{(\epsilon+\epsilon^4) z_1 + z_2}{\epsilon^2 - \epsilon^3}, \quad \frac{z_1 - (\epsilon+\epsilon^4) z_2}{\epsilon^2 - \epsilon^3}\right),$$

then
$$s^5 = 1, \quad t^4 = 1,$$

and s, t generate a group of 120 homogeneous substitutions, with which the group of rotations is hemihedrically isomorphic; because if $(\alpha z_1 + \beta z_2,\ \gamma z_1 + \delta z_2)$ is any one of the substitutions,

$$t^2(\alpha z_1 + \beta z_2,\ \gamma z_1 + \delta z_2) = (-\alpha z_1 - \beta z_2,\ -\gamma z_1 - \delta z_2)$$

which corresponds to the same rotation. Every one of the homogeneous substitutions leaves f, H, T absolutely unaltered, but produces a certain permutation among their roots.

Consider, now, the function T. It evidently has the rational factor

$$\phi_1 = z_1^2 + z_2^2;$$

and if we apply to this the substitution t, we find that $t(\phi_1) = -\phi_1$. Now the roots of $\phi_1 = 0$ are the ends of a diameter of the sphere: hence t must correspond to a rotation through an angle π about a perpendicular diameter, the extremities of which are unaltered by t, so that they are given by

$$\frac{(\epsilon+\epsilon^4) z_1 + z_2}{z_1 - (\epsilon+\epsilon^4) z_2} = \frac{z_1}{z_2},$$

or
$$\phi_2 = z_1^2 - 2(\epsilon+\epsilon^4) z_1 z_2 - z_2^2 = 0.$$

If we put
$$\phi_3 = z_1^2 - 2(\epsilon^2+\epsilon^3) z_1 z_2 - z_2^2$$

it is easily proved that the roots of $\phi_3 = 0$ are at the ends of a diameter perpendicular to each of the two others: hence, writing

$$\tau = \phi_1\phi_2\phi_3 = z_1^6 + 2z_1^5 z_2 - 5z_1^4 z_2^2 - 5z_1^2 z_2^4 - 2z_1 z_2^5 + z_2^6$$

τ is a factor of T and the roots of $\tau = 0$ are the vertices of a regular octahedron. Since this has 12 edges, there are 24 rotations which bring it into coincidence with itself; of these 12 belong to the icosahedral group, and form a factor of it.

By applying all the icosahedral substitutions to τ we obtain five different sextics $\tau_1 (= \tau)$, τ_2, τ_3, τ_4, τ_5 the product of which is T. If, now, we form the equation

$$(\tau - \tau_1)(\tau - \tau_2) \ldots (\tau - \tau_5) = \tau^5 + p_1\tau^4 + \ldots + p_5 = 0$$

the coefficients are binary forms which are invariable for the icosahedral group and of degrees 6, 12, 18, 24, 30 respectively. Each coefficient equated to zero must give an invariant set of points on the sphere; and since there are no sets of 6 or 18 points, and the only sets of 12 and 24 are given by $f = 0$, $f^2 = 0$, the equation must reduce to the form

$$\tau^5 + af\tau^3 + bf^2\tau - T = 0$$

where a, b are numerical. By a comparison of coefficients it is found that $a = -10$, $b = 45$, so that finally

$$\tau^5 - 10f\tau^3 + 45f^2\tau - T = 0.$$

Putting $$\tau^2/f = r$$

we find that r satisfies the equation

$$r(r^2 - 10r + 45)^2 = T^2/f^5.$$

The Hessian of τ is given by

$$\kappa = -(z_1^8 + z_2^8) + (z_1^7z_2 - z_1z_2^7) - 7(z_1^6z_2^2 + z_1^2z_2^6) - 7(z_1^5z_2^3 - z_1^3z_2^5);$$

like τ this has five conjugate values and is invariant for the same group as τ.

Suppose, now, that l, m are arbitrary numerical quantities, and let

$$y = \frac{lf\kappa}{H} + \frac{mf^3\tau\kappa}{HT} \quad \ldots\ldots\ldots\ldots\ldots\ldots\ldots\ldots(1).$$

This is a function of the ratio z_1/z_2 which assumes only five values when the icosahedral substitutions are applied to it. The invariant quintic of which it is a root can be found by a process similar to that by which the equation satisfied by τ was constructed. The result is, that if we write

$$\left.\begin{aligned} \frac{H^3}{f^5} &= j, \quad \frac{T^2}{f^5} = j_1 = 1728 - j \\ aj &= 8l^3 + l^2m + \frac{72lm^2 + m^3}{j_1} \\ bj &= -l^4 + \frac{18l^2m^2 + lm^3}{j_1} + \frac{27m^4}{j_1^2} \\ cj &= l^5 - \frac{10l^3m^2}{j_1} + \frac{45lm^4 + m^5}{j_1^2} \end{aligned}\right\} \quad \ldots\ldots\ldots\ldots\ldots(2)$$

y satisfies the equation

$$y^5 + 5ay^2 + 5by + c = 0 \ldots\ldots\ldots\ldots\ldots\ldots\ldots\ldots(3).$$

62. Conversely, suppose a quintic given in the form (3): if we can find, in terms of a, b, c, quantities l, m, j, j_1 such that $j+j_1=1728$, and the last three of equations (2) are satisfied, the roots of the given quintic will be expressible as rational functions of any one root of the normal equation

$$H^3 - jf^5 = 0.$$

By combining equations (2) we find

$$j_1(lb+c) = m^2 a \qquad\text{.................(4)},$$

$$j\left(lc - \frac{m^2 b}{j_1}\right) = \left(l^2 - \frac{3m^2}{j_1}\right)^3 \qquad\text{.................(5)},$$

$$j\left(\frac{la+8b}{m}\right) = l^3 + \frac{216l^2 m}{j_1} + \frac{9lm^2}{j_1} + \frac{216m^3}{j_1^2} \qquad\text{...........(6)}.$$

From (2) and (6), by squaring,

$$27a^2 j^2 = 1728l^6 + 432l^5 m + 27\left(1 + \frac{16.72}{j_1}\right)l^4 m^2$$
$$+ \frac{4320}{j_1}l^3 m^3 + 27\left(\frac{2}{j_1} + \frac{72^2}{j_1^2}\right)l^2 m^4 + \frac{18.216lm^5}{j_1^2} + \frac{27m^6}{j_1^2},$$

$$j_1 j^2\left(\frac{la+8b}{m}\right)^2 = j_1 l^6 + 432l^5 m + \left(18 + \frac{216^2}{j_1}\right)l^4 m^2 + \frac{4320}{j_1}l^3 m^3$$
$$+ \left(\frac{81}{j_1} + \frac{2.216^2}{j_1^2}\right)l^2 m^4 + \frac{18.216}{j_1^2}lm^5 + \frac{27.1728}{j_1^3}m^6.$$

On subtracting the last equation from the one before, we find that $(1728-j_1)$ is a factor of the right-hand side; since $j+j_1=1728$, this cancels with the factor j on the left hand, and we thus obtain

$$j\left[27a^2 - j_1\left(\frac{la+8b}{m}\right)^2\right] = \left(l^2 - \frac{3m^2}{j_1}\right)^3.$$

Comparing this with (5), we infer that

$$27a^2 - \frac{j_1}{m^2}(la+8b)^2 = lc - \frac{m^2 b}{j_1} \qquad\text{.................(7)},$$

and by eliminating m^2/j_1 from this and (4) it is found that l satisfies the equation

$$(a^4+abc-b^3)\,l^2 - (11a^3 b - ac^2 + 2b^2 c)\,l - (27a^3 c - 64a^2 b^2 + bc^2) = 0 \qquad\text{......(8)}.$$

If D is the discriminant of (3), that of (8) is $a^2 D/5^5$, so that l is rational in the field $(a, b, c, \sqrt{D}, \sqrt{5})$. The adjunction of $\sqrt{D}$ reduces the group of the quintic from the symmetrical group to the alternate group of order 60; the quantity $\sqrt{5}$ is what is called an auxiliary irrationality, and does not affect the group.

Having determined l, equations (4) and (5) give

$$j = \frac{(al^2 - 3bl - 3c)^3}{a^2\{(ac - b^2)\, l - bc\}} \quad \text{......................(9)},$$

a rational function of l; and since

$$\left(l^2 + \frac{m^2}{j_1}\right) m = aj - \frac{72lm^2}{j_1} - 8l^3,$$

we find, after substituting for j and m^2/j_1 from (4) and (9), that

$$m = \frac{a^2l^4 - 10abl^3 - (18ac - 45b^2)\, l^2 + 18bcl - 27c^2}{(ac - b^2)\, l - bc} \quad \text{...(10)}.$$

Thus m can also be expressed as a rational function of l: of course, the above expression, like that obtained for j, can be transformed in various ways by making use of the equation satisfied by l.

To make this method actually useful for solving numerical quintics, we require a table giving the roots of the icosahedral equation

$$H^3 - jf^5 = 0$$

for different numerical values of j. When D is positive, l, m, j are real; but when D is negative, j is in general complex, so that a complete table would have to include imaginary values of j.

63. When $a = 0$, the foregoing results require modification, because in this case $lb + c = 0$, and the formulæ (9) and (10) become indeterminate. Starting afresh with equations (2), after putting $a = 0$, it is found that if ξ is a determinate root of

$$b^2\xi^2 + c^2\xi - 64b^3 = 0 \quad \text{......................(11)}$$

we may put

$$\left.\begin{aligned} bl &= -c \\ b^4 m &= 72b^3c - c^3\xi \\ b^8 j &= b^3(1728b^5 + 63c^4) - c^2(c^4 + 81b^5)\,\xi \\ b^8 j_1 &= -63b^3c^4 + c^2(c^4 + 81b^5)\,\xi \end{aligned}\right\} \quad \text{.........(12)},$$

and the formula (1), combined with $H^3 - jf^5 = 0$, will give the roots of the equation

$$y^5 + 5by + c = 0.$$

Another special case that requires examination is when $a \lesseqgtr 0$, and equation (8) of last article is satisfied by putting

$$(ac - b^2)\, l = bc.$$

This leads to $3ac = 4b^2$, whence also, supposing that c does not vanish, $l = 4b/a = 3c/b$. It is found that the equations (2) of Art. 54 reduce to

$$j + j_1 = 1728,$$

$$j = \frac{2^{11}b^3}{a^4} + \frac{64b^2m}{3a^3},$$

$$\frac{m^2}{j_1} = \frac{16b^2}{3a^2}.$$

The elimination of j and j_1 leads to

$$9a^6m^2 + 2^{10}ab^4m + 3 \,.\, 2^{10}(32b^3 - 27a^4)\,b^2 = 0,$$

the roots of which are

$$-\frac{96b}{a}, \quad \frac{32b\,(27a^4 - 32b^3)}{9a^5}$$

and the corresponding values of j are

$$0, \quad \frac{2^{12}b^3}{27a^8}(27a^4 - 16b^3).$$

Now, if we take $j = 0$ the auxiliary equation is $H = 0$. Referring back to equation (1), Art. 54, we see that this must be rejected, because it introduces a zero factor into the denominator of the expression for y. Thus the solution is

$$y = \frac{4bf\kappa}{aH} + \frac{32b\,(27a^4 - 32b^3)\,f^3\tau\kappa}{9a^5HT},$$

with $$27a^8H^3 - 2^{12}b^3(27a^4 - 16b^3)\,f^5 = 0.$$

This may be simplified by putting

$$\frac{16b^3}{27a^4} = n:$$

thus $$y = \frac{4bf\kappa}{aH}\left\{1 + 24\,(1 - 2n)\frac{f^2\tau}{T}\right\},$$

with $$H^3 - 2^8 \,.\, 3^3n\,(n-1)\,f^5 = 0.$$

If $a = 2p^3$, $b = 3p^4$, this solution fails: but the equation is then

$$y^5 + 10p^3y^2 + 15p^4y + 6p^5 = 0;$$

that is to say,

$$(y + p)^3\,(y^2 - 3py + 6p^2) = 0,$$

the roots of which are obvious.

NOTES AND REFERENCES

8. The very important idea of a field of rationality has been made precise by Dedekind (Dirichlet-Dedekind, Vorlesungen über Zahlentheorie, Suppl. XI) and Kronecker (Grundzüge einer arithmetischen Theorie der algebraischen Grössen: *Journ. f. Math.* XCII = Werke 2).

15. On the problem of finding the irreducible factors of a polynomial, see Kronecker (*Grundz.*) and K. Runge (*J. f. Math.* XCIX). Special devices often shorten the work in particular cases.

Another way of finding the Galoisian group is explained by O. Hölder (*Encycl. d. math. Wiss.* I, p. 486). Except theoretically, the problem is not of much interest.

40. It will be observed that the definition of Abelian equations includes cyclical equations as a particular case; it is, however, convenient to retain both terms.

43, end. For the proof referred to, see Weber's *Algebra*, II, pp. 736–821 (or *Acta Math.* VIII), and Hilbert, Die Theorie der algebraischen Zahlkörper, chap. 23 (*Jahresb. d. deutschen Math.-Ver.* 1894–5).

47. Weber applies the term *metacyclic* to all groups for which the indices e_i (p. 21) are primes, and calls the corresponding equations metacyclic. Another (perhaps preferable) term is *soluble*. The definition of a metacyclic group given in the text agrees with that of Kronecker (*Berl. Ber.* 1879).

50. It was first satisfactorily proved by Kneser that, when the roots of a cubic equation are all real (the *irreducible case* in Cardan's solution), they cannot be expressed in terms of real radicals (*Math. Ann.* XLI. 1893).

62. The algebraical eliminations contained in this article appear to have been first carried out in this way by Gordan (see Klein, *Ikos.* p. 192, note).

The following list contains references to a selection of treatises and memoirs relating to the subject of this tract and its applications. Many more will be found in the *Encyclopädie der mathematischen Wissenschaften*, vol. I, sections B 1a, c, 3b, c, d.

L. Bianchi. *Lezioni sulla teoria dei gruppi di sostituzioni e delle equazioni algebriche secondo Galois;* Pisa, 1900.

C. Jordan. *Traité des substitutions et des équations algébriques;* Paris, 1870.

E. Netto. *Substitutionstheorie und ihre Anwendung auf die Algebra;* Leipzig, 1882 (trans. F. N. Cole, Ann Arbor, 1892): *Vorlesungen über Algebra;* Leipzig, 1896–9.

J. A. Serret. *Cours d'algèbre supérieure;* Paris (5th edition), 1885.

H. Weber. *Lehrbuch der Algebra;* Braunschweig, 1895–6 (2nd edition, 1898–9).

F. Klein. *Vorlesungen über das Ikosaeder und die Auflösung der Gleichungen vom fünften Grade;* Leipzig, 1884.

F. Klein and R. Fricke. *Vorlesungen über die Theorie der elliptischen Modulfunctionen;* Leipzig, 1890–2.

H. Weber. *Elliptische Functionen und algebraische Zahlen;* Braunschweig, 1891.

E. Galois. *Œuvres mathématiques;* ed. E. Picard, Paris, 1897 (also in *Liouv.* (1) XI).

N. H. Abel. *Œuvres;* ed. Sylow et Lie, Christiania, 1881.

P. Bachmann. *Die Lehre von der Kreistheilung;* Leipzig, 1872.

J. L. Lagrange (*Nouv. mém. de l'acad. roy. de Berlin*, 1770, 1771: or collected works, III).

C. Jordan (*Math. Ann.* I. 145, 583); L. Kronecker (*Crelle*, XCIV. 344, *Berl. Ber.* 1853, 1856); O. Hölder (*Math. Ann.* XXXIV. 26); E. Heine (*Crelle*, XLVIII. 237).

Full references to the literature of cyclotomy will be found in Bachmann's *Kreistheilung*: other special applications are

(1) The solution of the quintic, for which see R. F. A. Clebsch (*Math. Ann.* IV. 284); F. Brioschi (*Math. Ann.* XIII. 109); P. Gordan (*Math. Ann.* XIII. 375, XXVIII. 152); C. Hermite (*C. R.* XLVI. 508, and LXI–II *passim*); L. Kiepert (*Crelle*, LXXXVII. 114); L. Kronecker (*Crelle*, LIX. 306).

(2) The impossibility of expressing the roots of a general quintic in terms of radicals was not conceded by all (Jerrard being probably the last disputant) till 1863. An historical account of the subject is due to McClintock (*Amer. Journ.* VIII. 1886). Other papers on the quintic are by McClintock (*Amer. Journ.* XX. 1898); Cockle (*Phil. Mag.* 1854, 1863); Malfatti (*Siena Trans.* 1771); Runge (*Acta Math.* VII. 1885); Berwick (*Proc. Lond. Math. Soc.* XIV. 1915); Dickson (*Bull. Amer. Math. Soc.* XXXI. 1925).

(3) Polyhedral equations: H. A. Schwarz (*Crelle*, LXXXVII. 139).

(4) Equations connected with elliptic and modular functions: W. Dyck (*Math. Ann.* XVIII. 507); L. Kiepert (*Crelle*, LXXV. 255, and *Math. Ann.* XXVI, XXXII, etc.); Klein (*Math. Ann.* XII, XV); A. G. Greenhill (*Proc. Lond. Math. Soc.* XIX, XXI, XXVII, etc.).

(5) §§ 54–59. A fuller account of the resolvents of a sextic equation is given by Berwick (*Journ. Lond. Math. Soc.* III. 1928; *Proc. Lond. Math. Soc.* XXIX. 1929).

(6) Equation of seventh degree with simple group of order 168. F. Klein (*Math. Ann.* XIV. 428); P. Gordan (*Math. Ann.* XX. 515, XXV. 459).

(7) Determination of the inflexions of a plane cubic. O. Hesse (*Crelle*, XXXIV. 193).

Printed in the United States
By Bookmasters